Care and Repair of Shop Machines

Care and Repair
of Shop Machines

A Complete Guide to Setup, Troubleshooting,
and Maintenance

John White

The Taunton Press

The Taunton Press
Inspiration for hands-on living™

The Taunton Press, Inc., 63 South Main Street, PO Box 5506,
Newtown, CT 06470-5506
e-mail: tp@taunton.com
Distributed by Publishers Group West

COVER DESIGN: Lynne Phillips
INTERIOR DESIGN: Mary McKeon
LAYOUT: Rosalie Vaccaro
ILLUSTRATOR: Mario Ferro
PHOTOGRAPHER: Scott Phillips

LIBRARY OF CONGRESS CATALOGING-IN-PUBLICATION DATA:
White, John (John Wilson), 1949-
    Care and repair of shop machines : a complete guide to setup,
troubleshooting, and maintenance / John White.
        p. cm.
    ISBN 1-56158-424-X
  1. Machine-shop practice. 2. Tools--Maintenance and repair. I. Title.
    TJ1160 .W478 2002
    670.42'3--dc21   20002003444

The following manufacturers/names appearing in *Care and Repair of Shop Machines*
are trademarks: Biesemeyer®, Cool Blocks®, Craftsman®, DeWALT®, Dremel®,
DriCote®, Emerson Tool Company™, Fenner™, Fenner Drives™, Freud®, General®
Mfg Co. Ltd., Lenox®, Makita®, Milwaukee®, Panasonic®, Radio Shack™,
Rockler™ Woodworking and Hardware, Scotch-Brite™, Sears™, Starrett®, Sunhill
Machinery™, TopCote®, Woodcraft®, and Woodworker's Supply, Inc.®

ABOUT YOUR SAFETY

Working wood is inherently dangerous. Using hand or power tools improperly or
ignoring safety practices can lead to permanent injury or even death. Don't try to
perform operations you learn about here (or elsewhere) unless you're certain they are
safe for you. If something about an operation doesn't feel right, don't do it. Look for
another way. We want you to enjoy the craft, so please keep safety foremost in your
mind whenever you're in the shop.

Printed in the United States of America
10 9 8 7 6 5 4 3 2 1

To my parents, who encouraged me even when my aspirations exceeded my abilities

Acknowledgments

I was never a good student, so I never had a master as a teacher, except for my father, John White, who was a quiet master in many fields. Much of what I know, especially about machinery, came from the authors of *Popular Science* and *Popular Mechanics* who wrote in the 1940s and '50s. Only now, as I try to become a writer myself, can I appreciate their abilities both as craftsmen and writers.

Closer to home, I want to thank the editors and authors of *Fine Woodworking* magazine and the editors at Taunton books for their patience, encouragement, and editorial guidance.

Contents

Chapter 4 - The Thickness Planer   88

Chapter 5 - The Drill Press   120

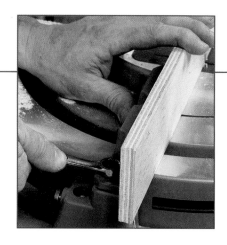

Introduction

There is great advantage in the consistent accuracy a woodworking machine can achieve. But this precision is not guaranteed, since no machine will cut square and straight forever. Even brand-new tools, straight out of the box, can't be assumed to be accurate. Some are not correctly tuned when they leave the factory. The parts are bolted together and brought approximately into line, but the final adjustments are always left up to the person using the machine. Some machines go out of alignment from having suffered bumps and shoves in their journey from the assembly line to your shop. In older machines, time and use take their toll. Parts wear, bolts may loosen, and even massive castings can warp.

The engineers and machinists who build power tools know that their machines will need adjusting, and all power tools are designed with this reality in mind. Some owner's manuals will supply the information you need to tune up a machine, but most manuals lack the detailed information you need to bring your tools into precise adjustment.

The purpose of this book is to supply the information you need to make your tools perform at their best, to pick up where the manuals, if they haven't been lost, left off. Although there have been dozens if not hundreds of models of each power tool built over the past century, most tools are very similar in their basic elements. The tune-up techniques in each chapter aren't specific to any one machine but will in most cases be adaptable to the power tools you own. Even if you don't see exactly your make and model machine in this book, you will be able to apply the basic principle.

There is no one right way to set up machinery. In fact, I've tried to offer alternatives where I could, especially low-tech or shopmade solutions in place of expensive, dedicated tools. If you own a professional shop and need to keep your machines in top form every day, then dedicated measuring tools and jigs may be right for you. If you're a home-shop woodworker, low-tech solutions are often adequate for the occasions when you check and maintain your machines.

Most important is to take the time to set up your machinery correctly and to maintain the settings. The result will be more accurate cuts, square, flat stock, and better, safer woodworking.

Tune-Up and Maintenance Tools

I f you are going to invest in woodworking machines, you will also need to invest in the tools to set them up and maintain them. A machine out of the box, brand new, is rarely in the best condition to perform accurately. If you're really lucky, your new machine may be fairly accurate,

A tune-up and maintenance kit includes familiar tools and some precision measurement tools.

but don't assume that it is. Check the beds and table for flat and fences and cutters for square. And then, of course, there may be annoying vibration or runout that can affect tool performance.

Think of your machines as kits. You get the basic parts, more or less in the right position, some instructions, and maybe even some tools to help you make adjustments. Unfortunately, this tool kit, if it happens to be included, is rudimentary. If you're going to fine-tune your machine, you'll need a few more things. If you've bought a used tool, you may be starting from scratch.

Fortunately, a tool kit for setting up and maintaining machinery needn't be extensive or expensive. If you work on your car or have a basic home-repair tool kit, you may already have some of the tools you'll need.

Tools Accompanying Your Machine

Although it might not seem obvious, information is a tool. Yes, you need wrenches and squares, but knowledge and information (including this book) are just as important to setting up and maintaining machines. The more you know about how your machine runs and its internal structure, the better prepared you will be to develop a workable approach for tune-up. Your most valuable tool may be the owner's manual the manufacturer provides with a new machine. These include helpful photos or drawings identifying the parts and explaining their roles in the operation of the machine.

Sometimes the mechanical connections are obvious simply from looking at the tool. But in some cases, it's impossible to see the working parts, never mind understanding how they interact. This is where the exploded drawings showing parts can be helpful. If you study these schematics, you

Getting the Owner's Manual

If you don't have the original owner's manual or schematics, these are sometimes available through a dealer selling the same brand of machinery. They can also be ordered directly from the manufacturer (see Sources on p. 198). Make sure you have your model number handy when you call to order.

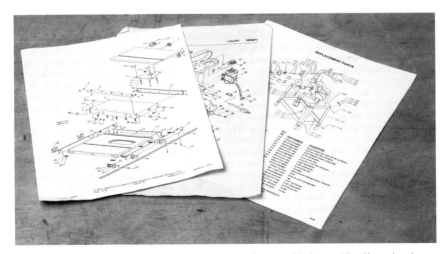

Not only can the schematic included with the owner's manual help you identify and order replacement parts, it also can help you understand how your machine works.

will get a good sense of how your machine is put together. These schematics are essential if you need to order replacement parts. Call-outs indicate individual parts, whose numbers can be looked up on a keyed parts list, so keep this information in a safe place.

Once you've familiarized yourself with the anatomy of the machines and the relationship of the parts, you should cull any information you can from the owner's manual itself. The information on setup and maintenance may be very basic, or even incomplete, but it's a start.

Other tools you may find in the box include wrenches that are specifically for the bolt and nut sizes of your machine. There may also be hex keys or specialty wrenches. Remember to check onboard the machine for tools; some manufacturers have now designed holders to keep setup tools nearby for quick adjustments. Unfortunately, manufacturer-provided tools are seldom the best quality, so if you have your own tools of those types and sizes, you'll probably want to use those.

In most cases, the manufacturer doesn't include all the tools you'll need to set up and tune up your machine. Measurement tools such as squares and rulers are never included, so you'll have to buy them. If you have a used machine, you are most likely on your own where tools are concerned. The previous owner may have included the owner's manual, but the original manufacturer's tool kits rarely come along. In either case, you'll want to gather the basic tools you need before you attempt a tune-up or maintenance. Let's look at the tools in a basic kit.

The Basic Kit

The basic tools you will need fall into two classes: mechanical tools, which are used for tightening, loosening, and adjusting, and measuring tools. In both cases, buy good-quality tools that will last a long time. Sets of wrenches, sockets, and hex keys will not only contain the sizes you need for your machine but also useful sizes for tuning up other machines and general mechanical work.

When choosing measuring tools, quality is even more important. How can you make your machine accurate using an inaccurate measuring device?

MECHANICAL TOOLS

The primary mechanical tools needed are wrenches. For most work, a set of ⅜-in.-drive socket wrenches and a set of combination wrenches will be all that is needed. The socket set should include a 3-in. and a 6-in. extension.

For set screws and adjusting screws, a set of hex keys is a necessity. For value for the dollar, it is hard to beat Craftsman® tools, especially when bought as sets for about one-third of the price of the tools if they were bought individually.

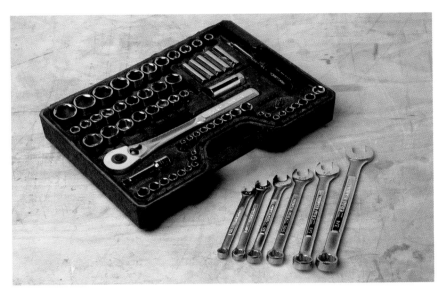

You need only a basic set of socket wrenches and combination wrenches to tune up most power tools.

If you are working on cast-iron American-made machines, you will need only inch-sized wrenches. The majority of the cast-iron woodworking machines made in Taiwan also use inch-sized hardware, but it isn't uncommon to find the odd metric-sized bolt thrown in. Most benchtop machines with cast-aluminum components are made with all metric-sized hardware no matter where the machine was made.

European and Japanese machines are consistently metric sized except for some older machines made in England, which will have British Standard nuts and bolts that are neither inch nor metric sized. You can always spot these machines because the bolt heads have been chewed up by someone using the wrong-sized wrenches. You can buy British Standard wrenches from Snap-On tool dealers and specialty catalogs for English sports car enthusiasts.

You will also need a set of screwdrivers, normally Phillips head, but you may have an occasional need for a slotted screwdriver or even a square drive. These needn't be especially good tools, but they do need to be sturdy enough to endure some torque. Especially on older machines, rust and other corrosion can make parts stubborn.

MEASURING TOOLS

To work their best, the critical components of a power tool need to be positioned and squared up with an accuracy of just a few thousandths of an inch. To a woodworker, a single thousandth (written as 0.001 in.) may seem vanishingly small, but machinists constantly work to this degree of precision. With just a few simple and inexpensive tools borrowed from the machine trade, you will be able to measure 0.001 in. as readily as you now measure 1/16 in. while woodworking.

Pulley set screws and many adjusting screws need hex keys to adjust them. A long-arm set where one leg of each wrench is extra length is especially useful.

By using a dial indicator, measuring a thousandth of an inch is as easy as reading a clock. The back of the indicator has a mounting lug with a ¼-in. hole drilled through it.

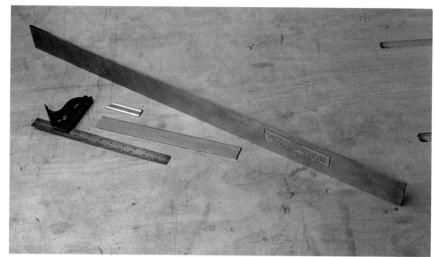

Machinist's straightedges, which are available in a variety of lengths, are useful for checking machined beds and tables for flatness.

For setting and checking the tables on a jointer, you need a feeler gauge set.

For checking surfaces for flatness, you will need an accurate straightedge. A hardware-store yardstick won't do. Aluminum rulers sold for construction in 4-ft. lengths are better but not as accurate as a machinist's straightedge. You won't find one of these at your local home center, but they are available from dealers that sell machinist's and engineer's tools.

The workhorse of precision measurement is a dial indicator. The plunger projecting from the side of the dial indicator causes the pointer on the tool's face to revolve. The pointer makes one revolution for each 1/10 in. the plunger moves; the 100 graduations around the perimeter of the dial each indicate a movement of 0.001 in. of the plunger. A second, smaller dial on the tool's face counts the number of full revolutions the main hand has made. The dial of the indicator can be rotated to bring the zero mark on the face in line with the pointer, a process called "zeroing" the tool.

The dial indicator used throughout this book and by far the most common style sold is a little more than 2 in. in diameter, has a range of 1-in. plunger travel, and reads in 0.001-in. increments. Chinese-made dial indicators of serviceable quality for the occasional tune-up are available for less than $20.

For tuning up a jointer, you will need a set of feeler gauges. A typical set for machine work contains 26 leaves ½ in. wide and 3 in. long. The thinnest leaf will typically be 0.0015 in., and the remaining leaves will start at 0.002 in. and increase in 0.001-in. increments to 0.025 in. Don't buy a smaller set sometimes sold for auto tune-ups; the smaller blades and fewer leaves in the set make them impractical for power-tool tune-ups.

The last item you will need is a good-quality combination square with a 12-in. blade. This is one tool where it is worth investing in a quality, machinist's-grade tool that will probably cost around $75. Avoid the temp-

A good-quality combination square is needed to fine-tune fence and blade alignments. To preserve its accuracy, the square shouldn't be used for woodworking.

tation of buying a less expensive square made for carpentry work because its accuracy is marginal at best. Take good care of the square and use it only for tune-ups and to check and adjust all of the other squares in your shop. If it is never used as a bench tool for day-to-day work, it should stay accurate forever.

A dedicated machinist's square is also an option, although it's less versatile since it's not adjustable and can't be used for measurement. Avoid dropping a square at all costs. Once it hits a concrete floor, a machinist's square is worthless for setup work and should be thrown away. All squares, whether for machine work or day-to-day woodworking, should always be protected from bumps and accidental dropping. A square will only stay square if it is treated with care and respect.

Commercial Jigs and Setup Aids

The message that machines need to be tuned must be getting out, since woodworking catalogs carry more setup tools than they once did. Now you can buy dial indicators from many of the better woodworking dealers, whereas not too long ago you'd have to go to a machinist's-supply dealer to buy one of these. But beside the basics, there are jigs, devices, and setup aids. Some of these are really useful and will save you some time; others are an unnecessary expense.

Whether you invest in these or not depends on many factors. How often will you need to retune your machines once you have them set to your satisfaction? If you have a cabinet saw, you may have to do this infrequently. But if you have a contractor's saw, which is notorious for getting out of alignment with the least use or bump, you may need to recheck the settings frequently. The other question is, what kind of work are you doing? If it's mostly furniture making, you probably want to set and keep very accurate settings. If you mostly do home improvement projects where accuracy is less important, regular maintenance may be enough.

I've tried to offer shop-made alternatives in this book wherever possible so that you don't have to spend money on commercial gadgets. All of the procedures I discuss require only a dial indicator, a set of feeler gauges, and a few easily made wood stands and straightedges. But let's look at some of the commercial devices.

SETUP JIGS AND FIXTURES

There's a lot of confusion in woodworking about what's a jig and what's a fixture. The old saying is that a fixture stands still, whereas a jig, well, does a jig and moves. Most of the jigs for setup do both. For example, the Super Bar holds a dial indicator fast, but it moves in the miter slot of a table saw to test whether the front of the blade and the back of the blade are in the same plane with respect to the miter slot.

This is a handy device, and its cost is not so great as to prohibit purchase. There are set screws to enable a tight fit in the miter slot, and it's designed for the standard distance most table saws have from the blade to the miter slot. If your saw has a broad distance between the blade and the slot, you may not find this device as helpful, since there is a limit to the adjustment. All that being said, the simple wood block I designed to hold a dial indicator for tuning a table saw works as well as the Super Bar and costs very little.

A dial indicator is very sensitive, and its readings off a surface will depend on the flatness and quality of the surface. Most people use the saw-blade itself as the reference when squaring the blade to the miter slot in table-saw tune-up. Remember that the blade may not be flat. If it's a

The Super Bar is designed to test whether the table-saw blade is aligned with the miter slot. A more accurate (and expensive) jig of this type is the A-Line-It System.

Originally designed to allow disk sanding on a table saw, this milled steel plate is also marketed as a setup aid.

brand-new blade, it may have a coating that will interfere with the reading. (Make sure you remove this coating before using the blade as a reference point.)

Because sawblades have this variability and the teeth or coating may interfere with the reading, toolmakers have created better reference points. One is a steel disk that is primarily designed to be a surface to which PSA abrasives can be attached. This turns your table saw into a disk sander. You can also use it instead of the blade for setup work. Keep in mind that this is not the primary purpose of this accessory, but it is more accurate a reference point than the blade itself.

A flat reference surface can be provided by a master plate (available from Woodcraft® and Hartville tools; see Sources on p. 198), which is a milled steel plate designed to be bolted onto the arbor of your table saw. This is a fairly expensive investment, which you may need only several times in your woodworking life, but it's a definite improvement over using the sawblade. The birch plywood plate, used the way I describe in the table-saw chapter, is more accurate than any of these steel tune-up plates and costs very little. I find it flat and dimensionally stable enough to do the job.

Most jigs are designed for table saws, since table saws are the workhorses of the shop and nearly everyone who does woodworking has one. But the next most vexing setup chore is setting jointer knives. Many of the most popular new jointers and planers feature knife setup that is self-setting. Older machines may require tedious fettling to get the blades square to the tables. To speed this and ease the tedium, several manufacturers have produced jointer-planer knife-setting jigs.

Again this is both a jig and a fixture. Magnets hold the jig dead-on the table surface, while the same force is used to pull the knives into alignment. Before you invest in one of these, make sure you need it. Modern self-setting systems obviously don't.

Magna Set is the most popular brand name in the jointer and planer knife-setting jig category. Before you buy one, try making the magnetic knife-setting jig I describe in chapter 2. I've found that it works better than commercial jigs and costs less than $10.

Making Your Own Setup Jigs

Woodworkers are an inventive group and enjoy making their own clever devices for every woodworking process. If this describes you, consider making some of your setup aids. These can be simple or elaborate. You need to make sure that whatever you use for jigs or tools has accurate reference surfaces. The most important thing to consider when using solid woods is the straightness of grain, quartersawing, proper drying, and avoiding reaction wood. Next look for dimensional stability. Here you have choices in man-made composites as well.

MDF

One of the modern wood composites is medium-density fiberboard (MDF). It used to be that this material was only available from dealers that served professionals. But MDF is so useful as a cabinet material, as a stock for moldings, and as a stable substrate for veneering that it's now more widely available. Your local home center probably stocks it. It's very heavy compared with plywood or solid wood and can be expensive, depending on where you purchase it and the quantity you purchase. But you don't need much of it to make jigs. Fortunately, home centers stock it in smaller sheets that are perfect for jig material.

MDF, a sheet good composed of wood dust and resin adhesives, is now available at many home centers.

Here are some shop-made setup tools constructed of birch plywood.

Remember that MDF is still a wood product and that wood moves. If you intend to use your jig beyond the initial setup chore, store it carefully to prevent it from changing its shape. Long pieces of MDF, such as straightedges, should be stored flat; if they're left leaned up against a wall for a few days they'll sag under their own weight and take on a permanent bow that will make them hard to recalibrate.

PLYWOOD

When cost is considered, cabinet-grade plywood is the next best choice for setup jigs. Baltic birch is better than MDF but at twice the price. Baltic birch plywood with many laminates is as dimensionally stable as MDF, if not more so. It is not typically available at home centers, so you may have to hunt for it among specialty dealers. This material (as does MDF) mills very much like wood, so it can be cut using woodworking machines. Make sure you remove any fuzz after the cut, so that your reference surfaces remain straight and true.

SOLID WOOD

If you use solid wood, make sure it mills cleanly and resists denting and compression. A good choice is a dimensionally stable hardwood. Maple obviously isn't a great option because it moves with every humidity and temperature change. Tropical hardwoods are very dimensionally stable but expensive. Some domestic hardwoods, including cherry, are better choices. In some cases, you'll just need a block of wood for distance. Since wood moves less in the linear direction, scraps of any wood species can be used for this purpose.

Other Useful Tools and Materials

There are some other tools and materials that you might find helpful in your setup and tune-up. Don't be afraid to be inventive; sometimes the solution is as simple as a shim cut from an aluminum can. It may not be conventional, but if it works, does it matter? The object is to get your machines to work for *you*.

SHAPING, GRINDING, AND FINISH REMOVAL

Sometimes you need to do some fine-tuning of metal parts. One device you may consider or may already have in your tool kit is a Dremel® tool.

Protecting Jigs and Setup Aids

IF YOU INTEND TO KEEP YOUR JIGS and setup aids for next time, make sure you prevent them from changing shape. MDF is very heavy and will sag under its own weight if stood against a wall, so make sure you store long jigs flat to prevent distortion. All wood-related materials will alter shape due to humidity changes over time. To be safe, check the accuracy of your jigs and setup aids after a period of storage by using references that you know to be straight and flat. If necessary, start from scratch and make a new jig rather than waste time trying to get a badly distorted one back into shape.

If you are someone who likes to finish jigs, think twice before applying finish to a jig for setup. In most cases, it's better not to finish them because a finish can interfere with the reference surfaces. Never use water-based finishes, which always swell the external surfaces and raise the grain.

A rotary tool can be equipped with a variety of small grinding wheels.

Keep a variety of shim stock handy and tin shears for sizing it.

This rotary tool can be fitted with any number of burrs and sanding drums to grind and polish. Remember that it is a rotary tool and will produce a rounded surface.

Manufacturers coat their tools with paint and finishes. The tools look good in the showroom but interfere with setup and accuracy. If the finish is in your way, get rid of it! The weapons here are grinding devices (stones and files) and abrasives (sandpaper), and in some cases chemical strippers.

Unless you handle sandpaper very carefully, it will almost always create a rippled or rounded surface on any metal. In descending order, this is how I would remove a finish from a machined surface on cast iron or steel:

1. If the finish chips off easily, which is fairly common on iron, I'd use a sharp chisel to scrape off the paint, taking care not to gouge the metal.

2. If the paint is tightly adhered, the only safe method is to use a stripper, but this is messy and risks damaging paint on adjacent surfaces.

3. Once the paint is removed, clean up the surfaces using a file laid flat on the metal. This will remove burrs and high spots without changing the flatness of the surface.

4. You can use a wire brush in a drill to remove paint on iron or steel, but the dust will spread all over the shop. On older machines, the dust will likely contain toxic metals, primarily lead and cadmium.

5. On soft metals such as die-cast zinc alloys, aluminum, and magnesium, the only safe method of paint removal is to use a stripper.

SHIMS

Shims are often necessary to achieve a tight or accurate fit of parts of machines. Metals make the best shims for this purpose because of their durability and stability. Consider copper or brass sheathing or aluminum, which is easily obtained from soda cans. To cut shims to size, you'll need shears or for small shims a wire cutter may do.

Choosing Sandpaper

For leveling or polishing machine parts, choose black wet-dry paper. It doesn't break down easily and it tolerates saturation with machine oil and other lubricants. Fine grits of wet-dry paper can be purchased at automotive-supply shops.

The Jointer

It is nearly impossible to build furniture with lumber that is not straight, flat, and square-edged, and the jointer is at the heart of careful stock preparation. The mechanics of a jointer are simple, but tuning and maintaining this tool can be challenging. A jointer requires very precise adjustments, within a few thousandths of an inch. When it is out of whack, a jointer will snipe the ends of boards, taper stock like a clapboard, create curved edges, and tear the surface.

Poor performance is most often due to several small misalignments, not just one. It is usually much more efficient to undertake a complete tune-up rather than try to isolate a single fault. This is especially true if the machine has never had a top-to-bottom check of its various parts.

Anatomy

A jointer has an infeed and an outfeed table, both of which typically can be adjusted for height (on some low-priced machines, the outfeed table is fixed). The tables flank the central cutterhead, which is mounted in the base casting (see the top left photo on p. 18). A fence along the back edge of the tables guides the stock, and a spring-loaded guard covers the cutting head.

On most American-made jointers as well as on their overseas knock-offs, the tables slide up and down sloped dovetailed ways in the base casting as they are adjusted in height (see the top right photo on p. 18). The dovetails need to be loose enough to slide but tight enough to hold everything in alignment. To achieve the proper fit, the dovetails are adjusted by set screws bearing on gibs (see the bottom photo on p. 18). On many

A properly tuned jointer will produce straight, square edges on stock, the first step in making quality furniture.

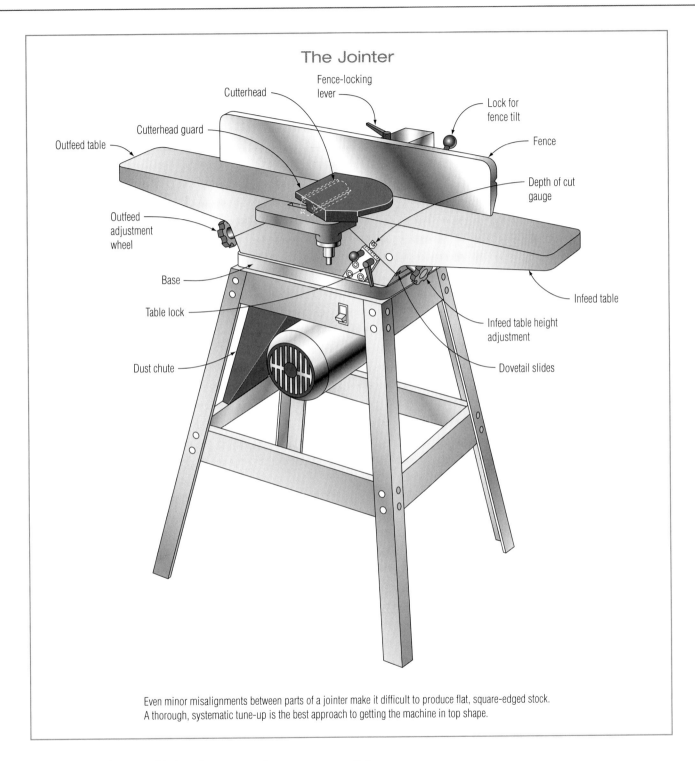

The Jointer

Fence-locking lever

Cutterhead

Lock for fence tilt

Cutterhead guard

Outfeed table

Fence

Depth of cut gauge

Outfeed adjustment wheel

Base

Infeed table

Table lock

Infeed table height adjustment

Dust chute

Dovetail slides

Even minor misalignments between parts of a jointer make it difficult to produce flat, square-edged stock. A thorough, systematic tune-up is the best approach to getting the machine in top shape.

European machines, table height is set with pivoting, parallel arms, an elegant design that works very smoothly and is less prone to problems.

When a board is jointed, the infeed table and the fence guide the stock as it crosses the cutterhead. The outfeed table picks up the freshly

Made from steel or aluminum, the cutter-head holds two or more replaceable knives.

Table height is set by sliding the tables on sloped ways. On this machine, a small T-handle locks the table, while turning the handwheel changes the height.

The end of the flat steel bar called a gib can be seen in this end view of the dovetail table slides. The gib-adjusting screw and its locknut is on the left; the bottom bolt and nut attach the jointer to its stand.

planed surface and guides and supports the stock as the pass is completed. The finished surface on the stock is only as straight as the path the wood takes across the cutter. If the tables aren't flat or if they do not line up with the cutterhead and each other, the path of the wood won't be straight and the finished edge of the board is bound to be crooked.

A full tune-up starts with cleaning and lubricating the jointer and checking the cutterhead bearings. Next, the tables are checked for flatness and the fit of the dovetail slides is adjusted. Once the slides are adjusted, the tables are shimmed if needed to line up with each other. The final part of the process is installing new knives in the cutterhead and fine-tuning the outfeed table height and the fence alignment.

You will need a set of feeler gauges, a small square, and a short straightedge (the blade from a combination square is ideal). If the tables need shimming, a 6-in. dial caliper would be handy for measuring shim stock, but the job can be done without one. To check the tables for flatness, you'll need long machinist's straightedges or shopmade master bars in two lengths. Later, when you are ready to change knives, you may want to make a magnetic jig to hold the knives in place as well as a very simple tool to locate top dead center for the cutterhead.

TROUBLESHOOTING

Because a jointer only has one function—to straighten the surface or edge of a board—the list of possible problems affecting performance is fairly short.

- The edge of a planed board is concave after repeated passes. In this case, the outfeed table is probably too low. If raising it doesn't correct the problem, the outfeed or infeed tables may be low at their outboard ends. Tightening the gibs normally solves the problem. The remaining possibilities for cutting a concave edge are worn slides and a warped table, both of which can often be corrected with shims.

- The jointer is cutting a convex edge. Here, the most likely cause is dull knives that are no longer removing as much stock as they used to. A second possibility is that the outfeed table is too high. A rare possibility is that one of the tables has warped upward or caked grease has somehow lifted one of the tables off of its slides.

Setting Up the Machine

For safety, the jointer should be unplugged during any adjustments or repairs. If you are planning to check and adjust the tables, remove both the fence and the blade-guard assemblies. On most jointers, this is a simple job. If you are just doing a blade change, the fence can stay in place but the guard should be removed.

Use a shop vacuum to remove loose sawdust from the jointer. Most of the dust and chips inside the machine won't be picked up by the vacuum, however, and should be dislodged by switching the hose on the vacuum to the exhaust side. I don't recommend using compressed air because dust and chips may be forced into the bearings and slides where they will eventually cause trouble.

CHECK THE BEARINGS

Cutterhead bearings and the drive pulley should be checked first, since a bad bearing must be replaced before the tune-up can go any further. Typically, replacing bearings on a jointer is not a difficult job.

Check the bearings by rotating the cutterhead a couple of revolutions. There should be no roughness, sticking, or noise. Next, lift up sharply on the pulley several times as you rotate the shaft by quarter turns (see the photo on p. 20). If you feel or hear a clicking sound, either the bearings are worn or loose in their housings or the pulley is loose and should be tightened.

Pay particular attention to the pulley. Small home-shop jointers seem especially prone to having it come loose. Check the pulley regularly because a loose pulley can quickly chew up the shaft. If the drag of the belt and motor makes it hard to judge the condition of the pulley or the bearings, remove the belt.

Always Turn Off the Machine

Although it seems impossible, a machine can inadvertently be turned on while you are working on it. To be absolutely sure the machine is turned off, always unplug any machine before attempting setup or maintenance operations.

By rotating and lifting the pulley, the jointer's bearings and pulley attachment can be checked for wear and looseness.

LUBRICATION

If the table-height adjusters are hard to turn, it is usually because the grease on the dovetails has dried out. Before taking the jointer apart for regreasing—a major job—try spraying a penetrating oil such as WD-40 onto the slides as you work them back and forth with the adjusting screws (see the top photo on the facing page). With luck, the oil will mix with the grease and revive it. You should also spray oil on the threads of the adjusting screws, reaching through the gap under the tables behind the adjusting handles.

If the height adjusters are still hard to turn after applying oil, either the gibs are too tight or the dovetails are clogged with dirt and rust. Try to adjust the gibs as explained in the next section. If that doesn't solve the problem, the tables must be removed so they can be cleaned and the slides greased.

Removing Tables for Maintenance

To clean and grease a jointer, slide the tables off the dovetails, but keep in mind that cast iron is heavy. A table on an 8-in. jointer can easily weigh more than 100 lb. You should have an assistant when you remove or install the tables on all but the smallest machines. Tackle one table at a time and reinstall it before moving on to the second one.

Using a penetrating oil to free up a jointer's tables is an effective, although temporary, alternative to taking the machine apart to clean and regrease it.

Choosing Lubricants

Penetrating oil is a temporary fix at best because it evaporates and gets absorbed by dust after a few days. Old grease doesn't lubricate as well as fresh grease, risking accelerated wear of the dovetails. You can improve the situation somewhat by trying to work in some thicker oil. But eventually, all jointers should be disassembled for a thorough cleaning and regreasing.

Before disconnecting a table's height-adjusting screw, the gibs should be locked to prevent the table from sliding free and crashing to the floor.

REMOVING TABLE EXTENSIONS

Begin by removing any bolted-on table extensions that support the guard or fence. Next, on most machines, the block that the height-adjusting screw threads into must be removed. The block is bolted to the base casting and projects up into the wedge-shaped area under the table.

Once the last gib screw is loosened, the table slides off the dovetailed ways that hold it to the base casting. Even on a 6-in. jointer, the table can be too heavy for one person to handle.

The final step before removing the table is to loosen the top and bottom gib-adjusting screws. Now only the center screw is holding the table in place.

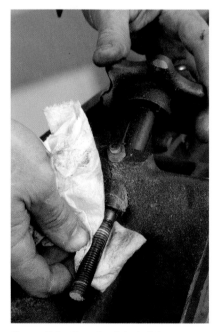

Clean off the threads of the height-adjusting screw. The threaded block may need soaking in paint thinner to remove old grease and debris.

Before unbolting the block, you must make sure the table won't slide by firmly tightening down the locking lever (see the bottom photo on p. 21). On machines that don't have a locking lever, tighten the center gib screw. This is important for your safety: If the table isn't locked, it will slide free and drop to the floor the moment the block is unbolted. On some jointers, it is possible that the weight of the remaining table will tip the machine over when the opposite table is removed. For safety, have a support under the table you are not removing.

Bolt heads holding the block in place are on the underside of the base casting and can be difficult to reach. In some cases, the stand will block your access. Typically there are two hex-head bolts holding the block but they may be Allen heads. Once the block is unbolted, you can remove the table.

Leaving the center or lock bolt for last, back off the top and bottom gib screws a couple of turns (see the photo above). Now only the center screw or lock bolt is holding the table in place. Steady the table and release this last bolt, then slide the table down and off the dovetails (see the top photo at left). On some machines you may have to slide the table up over the cutterhead to remove it. Set the table, dovetails up, on a pad of plywood on the other table of the jointer, making sure the machine won't tip over from the weight.

Use paint thinner and rags to remove old grease on the jointer's slides. A brush or Scotch-Brite™ pad will help on stubborn spots. In addition, remove the threaded block from the adjustment screw and clean up the adjustment screw and the gib (see the photo at left). Once everything is

Lubricate both ends of the shaft of the adjusting screw where it goes through the bushing in the table.

Using a small, disposable brush, apply grease to all sliding surfaces of the slides, a total of eight faces: four on the table and four on the base casting.

Filing Gib Surfaces

GIB SURFACES ARE OFTEN IN POOR CONDITION. To remove high spots without changing the gib's shape, I use a small, flat file on the gib, holding it flat on the surface (see the photo at left below). Use the file to bevel the edges at each end of the gib to make it slide more easily, and use a triangular file on the bearing surfaces of the dovetails (see the photo at right below). Keep the file flat on the machined surfaces of the casting and bear lightly: You just want to polish things up and remove burrs.

The faces of the gib often are burred or uneven from the pressure of the adjusting screws. Gibs should be dressed with a flat file and have its edges beveled.

Use a triangular file to pass lightly over both the flat and angled surfaces of the dovetails to remove burrs and rough spots.

There is only one correct orientation for the gib, where the tips of the screws match the indentations in the side of the gib. Before sliding the table back in place, determine how the gib should be installed.

clean, I always go over all the sliding surfaces with a file to remove burrs and rough spots.

Lubricate the height-adjusting screw where it goes through the bushing in the table with light oil, applying it at both ends of the bushing (see the top left photo). White lithium grease is best for the sliding surfaces of the dovetails. It can be applied with a small brush (small, disposable brushes called flux or acid brushes are perfect). Be sure to grease both the flat and the undercut beveled surfaces of the dovetails on the table and then grease the height-adjusting screw.

Once the table is done, grease the slides on the base casting (see the top right photo on p. 23). On some jointers, only the ends of the slide are bearing surfaces; the middle section is machined slightly lower and doesn't need greasing. Don't forget to grease the inside surface of the dovetail.

REASSEMBLY

Begin reassembly by threading the adjusting block onto the adjusting screw. Run it halfway down the thread. Match up the indentations on the back of the gib with the gib screws to be sure you will be sliding the gib back in its proper orientation (see the bottom right photo on p. 23). Some gibs have relief cuts that go against the dovetail surface; they are not pockets for the tips of the gib screws (see the photos below). Note the proper orientation of the gib and put it where you can reach it easily after you slide the table back in place. As long as you greased the dovetail surfaces well, you don't have to grease the gib. It will be easier to handle if you don't.

It is important to have a helper when reassembling the table. As you flip the table over, make sure the threaded block is rotated to its correct position. Carefully fit the dovetails together and slide the table back in place (see the photo at left). You may have to reach under the table to position the threaded block if it catches on the base casting.

After you have slid the dovetails all the way home, hold the table in place and slip the gib into place, being sure to have the correct side facing

Slide the table back onto the base with care to prevent jamming the dovetails. On larger machines, having a helper is a must.

The sharpened tips of the gib screws typically make small indentations in the gib. When reassembling the jointer, make sure the screws engage these marks.

On some gibs, a small pocket is milled in the gib on the face opposite the adjusting screws. This prevents the pressure of the screw from creating a high spot that would cause the gib to dig into the dovetail. Don't mistakenly install the gib with the pockets facing the screws.

the gib screws. On most machines, the end of the gib should come flush with the end of the base casting. You can check the gib's position by removing one of the adjusting screws and peering through its threaded hole in the base casting. If the gib is properly positioned, you will see the indentation from the screw point centered in the hole.

Once the gib is in place, lock the table by tightening the center lock handle or the center gib screw. For added security, tighten the top and bottom gib screws until they are snug against the dovetails. It should now be safe to release your grip on the table, but be careful when you let go.

With the table secure, turn the height-adjusting screw to position the block over its mounting holes in the base. Reaching up from underneath, thread the bolts back into the block and tighten them securely (this is often easier said than done). With the machine greased and reassembled, it's time to move on to a much more pleasant job: adjusting the gibs.

Gib Tune-Up

Gib adjustment is critical. When gib screws are loose, the tables drop at their outboard ends, throwing them out of alignment and making it impossible to create a straight edge on a board. Gib screws are not tightened equally. The screw nearest the cutterhead is tightened first. It is adjusted to apply the most pressure because the top end of the gib must counteract the weight of the table.

SCREW ADJUSTMENT

Begin by backing off all the gib screws a half turn to a full turn to leave the gib slightly loose but not so loose that the gib slides out of position. The table should now move very easily with the table-height adjuster. If the table binds, you should track down the cause and correct it now. Tighten the screw nearest the cutterhead while running the table back and forth with the height-adjusting knob (see the photo at left on p. 26). As you tighten the screw, the table will become harder to move. When this first screw is properly adjusted, it should take moderate effort to move the table. Once the adjustment feels right, hold the screw against turning and snug up the locknut (see the photo at right on p. 26). Tightening the locknut may change the pressure on the gib, so you should check the setting and readjust it if necessary.

Once you have the first screw adjusted and locked, repeat the procedure with the screw on the opposite end of the gib, tightening it to add only slightly to the force it takes to move the table. If your machine has a center screw with a locknut, adjust it last and with only light pressure against the slide. If the center screw is fitted with a knob so the table can be locked during regular operations, just leave it loose for now.

Tighten the top gib screw first because the leverage from the table is concentrated entirely on this end of the gib.

When the pressure on the gib feels right, lock the gib screw's lock-nut and then recheck the table's motion. Tightening the nut may affect the setting.

After you finish adjusting the gibs on one table, repeat the procedure on the second table. When all the gib screws are set correctly, you should not feel that you are straining either the machine or yourself to move the tables, but there should be no free play in the gibs. Getting the gibs adjusted just right is a matter of both technique and feel, much like tuning a musical instrument.

Adjusting Tables for Flatness

It may come as a surprise, but cast iron warps easily, typically because the casting process was faulty or because the rough castings were not allowed to age long enough before they were machined. Like wood, cast-iron tables can cup, bow, and twist. And just as with wood, checking for warp is done using a straightedge laid across the table's surface. The accuracy required is well beyond the tolerances of woodworking, however. You will need a straightedge that is accurate to within a couple of thousandths of an inch over its length.

I adapted a machinist's technique for creating a straightedge from inexpensive materials. Technically, the tool isn't a straightedge because only three slightly proud screw heads along one edge are precisely in line. More properly, the tool is called a master bar.

MASTER BARS

Although it may not seem possible, three screw heads along the length of a master bar can be aligned with each other so they are within thousandths of an inch of being in a straight line, creating an inexpensive straightedge whose accuracy rivals a costly machinist's tool.

A master bar is made from MDF, which is readily available and inexpensive. A master bar is actually made as a three-bar set in which one bar is designated as the master bar. The other two bars are nearly identical and are called setup bars. The setup bars are marked with the letters A and B.

PREPARING THE STOCK

The bars shown are made from ¾-in.-thick MDF. Foot-wide bull-nose MDF shelving is convenient; it is easy to rip a piece into narrower strips on a table saw. Cabinet-grade plywood also can be used, provided it does not have any bow that would prevent it from lying flat on a benchtop while the screws are being adjusted. Plywood has one advantage: It's lighter.

Bars up to 3 ft. long can be ripped to 4 in. wide. Bars up to 6 ft. long should be 6 in. wide. If you need an 8-ft. bar, it should be at least 9 in. wide and made from plywood for ease of handling. After cutting the bars to size, round off their sharp edges to make them more comfortable to work with.

For checking the flatness of a machine's tables, make the bars as long as each table. For checking the overall alignment of the jointer, you will need a bar that spans the full length of the machine.

All three bars in the set are cut to the exact same size. In each set, cut the top corners off of one bar to make it easier to spot as the master bar (this is the only difference between the master bar and the setup bars). On the machine-length bars, cut the top corners of the master bar back with a long taper to reduce the bar's weight.

Each bar has three 1⅝-in.-long fine-thread drywall screws set in one edge. The outer screws are placed 1 in. from the ends. On the shorter bars for checking table flatness, the third screw is placed at the bar's midpoint. On the machine-length bars, the third screw is located so it falls on the infeed table 1 in. from the cutterhead opening. Two 1-in.-dia. finger holes are drilled midway between the screws, close to the working edge of the bar.

MAKING THE BARS

After cutting the stock for the bars to size, clamp the three bars together and use a square to mark the center lines of the screw holes on all three bars at once (see the top photo at right). Use a fairly large drill bit to make pilot holes for the screws to prevent the edges from splitting. It's a good idea to test the bit and screw combination on a piece of scrap.

For the bars to work properly, it is important that the screw holes be square to the edge of the bars. Since it would be difficult to set up the bars on a drill press, make the jig shown in the top photo on p. 28 to guide the drill. For accuracy, use a drill press to make the hole in the jig's hardwood guide block before attaching it to the jig's larger fence piece. You can hold the jig by hand against the side of the bar while drilling the holes (see the photo at right).

To make sure that screw spacing is the same on all three bars in a set, clamp the bars together and lay out the holes on all three at once.

Line up the mark on the side of the jig with the line made earlier across the edge of the bar using a square. This ensures that the hole is precisely located.

A hole drilled in the hardwood strip on the side of this jig keeps the bit square to the edge of the bars when drilling pilot holes for the drywall screws.

To check that screw heads are even, stand them point up on a smooth surface and take a close look.

A crooked screw head also would throw off the bar's accuracy. Remove any burrs on the screw heads with a few passes on a file or sharpening stone and then stand the screws on their heads on a smooth surface such as a scrap of MDF (see the bottom photo above). Look at the screws from the side to see if they are straight, sighting across the upright edge of a square if you don't trust your eye. Look at the screws from both straight on and from one side to check for square in both directions.

The screw heads should all stand proud of the bar by about ¼ in. A simple way to set the screw height is to take a scrap of wood and cut a ¼-in.-deep notch in one corner. Set the block on the edge of the bar and adjust each screw in or out until the face of the notch just passes over the screw's head (see the photo at right). Once you have three bars with three screws in each, mark the two setup bars with the letters A and B and you are ready to fine-tune the screw height.

Check each table by placing the master bar along the table's back edge and then along its front edge. If the tables are bowed, use a feeler gauge to measure the gap under the bar (see the photo below). A gap of up to 0.005 in. is acceptable on a mid-sized machine. Beyond that, you'll start having trouble getting really straight edges on the boards you joint. After checking along the front and back edges, place the master bar on the table diagonally. Here again the gap should only be a few thousandths of an inch.

There is no foolproof or easy fix for a table that is out of flat. If the machine is still under warranty, the best course is to contact the manufacturer. Having the tables remachined is possible, but it is likely to be very expensive. But warped cast iron can sometimes be shocked back into alignment. To do this, remove the table from the machine and prop it up between blocks on the floor, then stand on the casting and give a small jump. If you are lucky, it will straighten out. If not, try again with a slightly bigger jump. Be careful: Too much enthusiasm can crack the casting and you can twist an ankle. This technique works more often than you might suppose, and it is actually recommended by one manufacturer for straightening warped jointer fences.

A block with a ¼-in. notch in one corner makes it easy to adjust all the screws to project equally.

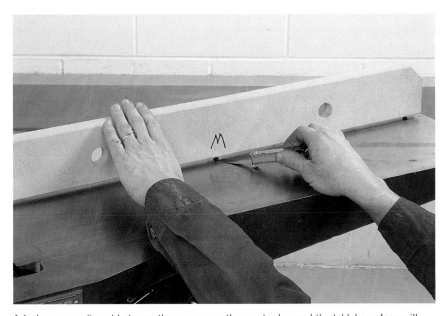

A feeler gauge slipped between the screws on the master bar and the table's surface will give you an accurate measure of how much the table is warped. It should be no more than a few thousandths of an inch out of flat.

Adjusting a Master Bar

ADJUSTING A SET OF BARS TO GET THE SCREW HEADS INTO A perfectly straight line is a four-step procedure that you will repeat two or three times. Since each step requires a slight adjustment of one or two screws, the entire process will take just a few minutes.

The only finesse required is in judging whether there is a slight gap between the center screws when you pair up a set of bars. If the screws are high, the bars will rock slightly as you alternately pull the bars together with the finger holes on either end of the bar. Even if the screws are high by just a thousandth of an inch, the rocking can be seen, felt, and even heard because the outer screws will click as they're brought together.

By comparison, the only clue to a gap between the center screws is visual, so it helps to have a good light and a light-colored background under the center screws as you make the adjustments. If you are making an adjustment to close a gap, the most efficient approach is to overshoot the adjustment slightly to make the bars rock and then back off the screw being adjusted until the rocking just disappears.

To get a feel for how the procedure works, start the first adjustment with the center screw on the master bar slightly low. If you set the initial screw heights with a notched block, the screws may already be quite close to being in line, so turn the center screw on the master bar in one turn to be sure it is low. Once you have a feel for the procedure, you can start with the center screw in the master bar at any height. Note that in all cases only the height of the center screws are changed when adjusting a set of bars.

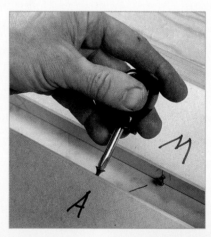

1 Place bar A and the master bar together and adjust the center screw on bar A to make all three pairs of screws touch.

2 Place bar B and the master bar together and adjust the center screw on bar B to make all three pairs of screws touch.

3 Place bar A and bar B together and equally adjust the center screws on both, both in or both out, the same number of turns to make all three pairs of screws touch.

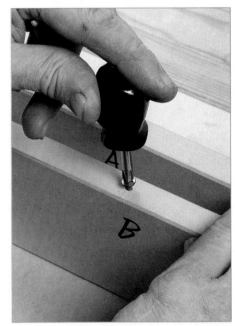

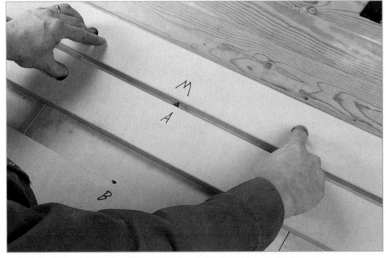

4 Place bar A and the master bar back together. This time adjust the master bar's center screw until all three pairs of screws touch. Do not adjust the screw in bar A.

Repeat steps 2, 3, and 4 until you find that there is no need to make any adjustments. At that point the screws on all three bars are in nearly perfect alignment and the master bar is ready to use.

The first step in checking the crosswise alignment of the jointer is to bring the tables level with each other along the back edge of the machine.

To check whether a straightedge is actually touching both tables evenly, try to slide a very thin feeler gauge, 0.001 in. or 0.002 in. thick, beneath the straightedge along its length.

Shim Stock

Brass is the best material for shimming the dovetails because it will not tear or compress as aluminum can and it won't dig into the cast iron as steel shims will. Assortments of small brass shims are available at most hobby shops for just a few dollars. Machine-shop suppliers also sell brass shim-stock assortments. The aluminum in soda cans is relatively hard and can serve as shim stock, but it is of limited use because it comes in only one thickness, normally around 0.005 in.

CROSSWISE TABLE ALIGNMENT

The next step is to test the crosswise alignment of the infeed and outfeed tables where they flank the cutterhead. If the tables don't line up crosswise, facing wide boards will create a wedge effect noticeable after a few passes. Your stock will start looking like a clapboard—thinner along one edge than the other.

Begin by raising the infeed table until a 12-in. metal straightedge laid on that table clears the cutting head when the knives are rotated out of the way. Now place the straightedge close to the back edge of the infeed table and extend it over the outfeed table by a few inches (see the photo at left above). Raise the outfeed table until it just touches the straightedge. As a quick check that the straightedge is evenly touching both tables, run the thinnest blade of the feeler gauge set along the length of the straightedge where it contacts the tables. It should not be able to slip under the edge anywhere along its length (see the photo at right above).

Next, move the straightedge to the front edge of the tables and check if it still touches both tables. If there is no gap, the tables are aligned. If there is a gap of more than a couple of thousandths of an inch when measured with a feeler gauge, the tables need to be brought into alignment by shimming.

SHIMMING

Shims are used to fine-tune the relative positions of the parts of a machine. They are typically placed between parts that do not move once they are bolted into place, but this is not the case with jointer tables. Here, shims must be placed between the sliding surfaces of the table and the base casting. Although you could shim either table to correct a misalignment, shimming the outfeed table is the better option because it is moved less frequently, so shims are less likely to wear out or work loose.

If the outfeed table is cast in one piece with the base, you have no choice but to shim the infeed table. A third possibility is that the outfeed table is a separate casting bolted onto the base. This style of machine can be shimmed by loosening the table's mounting bolts and slipping the shims between the base and table.

To avoid twisting the tables, shims are always used in pairs of equal thickness. To adjust the tilt across the table, place the shims at both the top and bottom ends of one slide. To adjust the lengthwise tilt of a table, place the shims at the top end or bottom end of both slides.

Shims should be an inch or two long, around one-sixth of the total length of the slide. You only need to pick up the ends of the slide to support the table properly. The width of the shim should be enough to cover the width of the slide with an additional ½ in. sticking out the side for easy handling (see the photo at left below). This tab can be folded down against the side of the machine once you are done, allowing a quick visual check that the shim hasn't shifted out of position (see the photo at right below). You can easily cut aluminum or brass shim stock with a good pair of scissors.

To adjust the crosswise alignment of the tables, cut two shims out of stock that is close in thickness to the amount of misalignment you measured with the feeler gauge. To insert the shims, back off the table's gib screws a turn, lift up on the low edge of the table, and slip the shims into place. Before installing the shims, give them a light coat of grease for lubrication and to keep them from slipping out of position while the gib screws are being adjusted. Once the shims are in, adjust the gib screws for a good sliding fit and recheck the table's alignment across the cutter opening.

Shims on a jointer slip between the table slides and the machine's base. Cutting the shim oversize leaves a tab that's useful in the positioning process.

Once the tables are aligned, the projecting edge of the shim can be folded up against the side of the table. Later, the tab will make it easy to check whether the shim has moved.

You may have to experiment with shim thickness a few times before getting it right.

LENGTHWISE TABLE ALIGNMENT

Because of the tables' weight and long overhang, it is fairly common to find that the ends of the tables have dropped, causing the machine to be high over the cutterhead. This condition will cause a concave cut in the edge of boards when they are jointed. Before checking the machine's alignment with a master bar as described next, always make sure the gibs are tight.

To check the lengthwise alignment of the tables, you'll need to make a long master bar using the method described on pp. 26–31. This large bar is 6 in. wide and as long as the jointer. There are screws in 1 in. from either end and one near the middle that touches the infeed table 1 in. back from the cutterhead opening.

To check the lengthwise alignment of the jointer, bring the tables up to the same level using a short straightedge across the cutter opening. Most of the straightedge should bear on the infeed table with only 1 in. or so

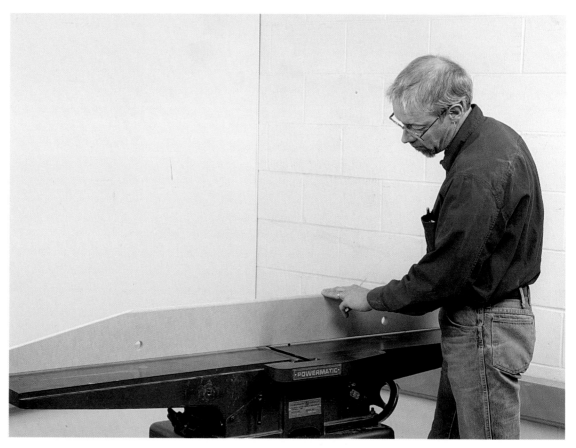

It is fairly common for jointer tables, which have a lot of overhanging weight, to sag at their outboard ends. A master bar as long as the machine is used to check whether tables are in the same plane.

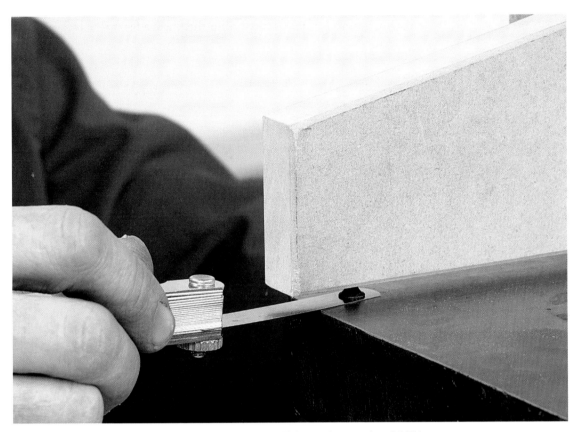

Using a feeler gauge, check the gap under the bar screws. If the gap is more than 0.005 in., shim the table to bring it back in line.

extended over the outfeed table. Raise the outfeed table so it just touches the straightedge. The two tables should line up across their full width (if not, they should be aligned crosswise as outlined in the previous section).

Next, place the long master bar across both tables, close to the back edge, using two screws on the infeed table and one on the outfeed (see the photo on the facing page). Ideally, the tables will form one continuous, flat surface and the screws will bear equally along the length of the master bar. If there is a gap under a screw head, use a feeler gauge to measure it (see the photo above). After checking using the master bar along the back edge of the machine, repeat the check along the front edge of the tables.

If the same gap appears when the bar is used at the front and back of the tables and the gibs are properly adjusted, then one table should be adjusted using a pair of shims at either the top or bottom of its slides to bring the tables into a common plane. The basic procedure is the same as was outlined for adjusting the crosswise table alignment, except that both shims are placed at the top or the bottom of the table slides.

Because of the overhang, a shim of 0.001 in. placed in the dovetail will typically raise or lower the far end of the table approximately 0.003 in.

Try to get the tables aligned to reduce the gap under the screw on the master bar to 0.005 in. or less.

If the table already has a pair of shims from adjusting the crosswise alignment, you can stack the shim for the lengthwise adjustment on top of the existing shim. Sometimes the stacked shims work loose when the jointer is back in service, so either replace the stack with a single shim of greater thickness or glue the shims together with a drop of Super Glue.

If the there is a substantial difference in the gap under the screw head between the front and rear edges, recheck the tables for crosswise alignment and flatness, looking especially for twist. To align a jointer whose tables are twisted, concentrate on aligning the back portion of the tables along the fence. This will give you one decently straight path through the machine for edge jointing.

Fence Alignment

Before reattaching the fence, check it with a straightedge. A small crown or dip over the length of the fence can be tolerated as long as all parts of the fence can be kept vertical to the tables. A twist or wind, to use the woodworking term, creates a problem though, because it will cause the wood to rotate as it passes down the length of the fence. The same solutions for straightening an out-of-flat table apply to a warped fence, with the one additional option of attaching a new wooden face to the casting and either shimming or planing it to a true surface.

Once the fence is remounted on the jointer, square it up and take a test pass with a board that has a good, flat face. Check the resulting edge using a reliable square, being sure to place the stock of the square against the board surface that ran along the fence. Adjust the fence stop as needed to get a square edge on the board.

Setting and Replacing Knives

Each jointer knife must be carefully adjusted so its cutting edge is even with the plane of the outfeed table as the knife passes through the highest point in its arc. When all the knives are set correctly, the freshly cut surface of the board will be supported and guided by the outfeed table, creating a perfectly straight edge.

Adjusting jointer knives can be a frustrating experience. On most home-shop machines, height adjustments are made by moving the knives up or down in slots milled into the cutterhead and then locking them into place. The height of the knife is almost always measured by some tool standing on the outfeed table. The tool can be as simple as a block of wood extending over the cutterhead or as complicated as a dial indicator on a customized stand.

MAKING A PAIR OF MAGNETIC jigs is simple. Each jig has three powerful bar magnets about 1 in. by 2 in. by ½ in., which you can purchase at Radio Shack™. The piece the magnets are glued to is made from ¾-in. plywood about 10 in. long.

For each jig, line up three of the magnets on your jointer's table as shown in the top photo, with two of the magnets near the ends of the jig and the third magnet spaced ¼ in. away from one of the end magnets. Make sure there are no metal chips stuck between the face of the magnets and the table. Using silicone adhesive, glue the plywood to the back of the magnets (see the center and bottom photos). Allow the glue to set overnight before moving the jig.

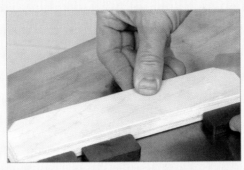

Use the piece of plywood to line up the magnets on the jointer's table. Make sure that there are no metal chips caught under the magnets.

Apply silicone adhesive to the magnets. Epoxy will work, but the bond will be weaker.

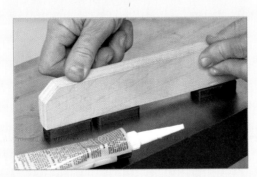

Press the plywood backing into the bead of adhesive. The jig should set overnight before being moved.

With disposable knives, always take care to remove all gum and pitch from the knife, the slot, and the knife-locking wedge because clearances are tight on this type of cutterhead.

Most methods for setting knife height are fussy and time consuming but still only moderately accurate. Compounding the problem, a knife often will shift out of position as its lock screws are tightened. This section will tell you how to solve both problems. Cleaning up and removing burrs from the knife-lock components will solve the shifting problem; building a simple magnetic jig will make positioning each knife a simple task (see the sidebar on p. 37).

On the simplest style of cutterhead, there is no way to adjust the knife position except to nudge the blade up and down, usually with the tip of an awl slipped in behind the knife from the side of the machine, a slow and tedious job at best. With this type of head, a magnetic jig is almost a necessity for changing a set of knives.

Two improvements to the hunt-and-peck approach to knife setting are spring-backed knives and cutterheads with height-adjusting screws behind the knives. Although both methods work, neither is especially easy or foolproof. Spring-backed knives are pushed down into their slots against a pair of small coil springs set into holes in the bottom of the slots. Typically the knife is pushed down by a wood straightedge placed on the outfeed table. Holding the knife against the springs while tightening the lock bolts can be awkward and requires adjusting each end of the knife repeatedly until it is seated properly. A second problem is that the springs pack up with sawdust. They are quite small, and a fumble when removing them for cleaning sends them down the dust chute or into some hidden corner of the jointer's base casting.

On heads with blade-height-adjusting screws, a pair of small stepped blocks, with the adjusting screws threaded into them, ride in grooves behind the knife slot. Turning the Allen-head screws lifts the knife higher in the slot. Exposed to pitch and dust, the adjusters and their grooves must be thoroughly cleaned with every knife change.

Setting knife height with adjusters requires either working with a dial indicator on a stand or bringing the blade up against a straightedge laid on the outfeed table while rocking the head back and forth. Either method of judging the height requires careful attention and repeated checks at either end of the knife while tweaking the adjusting screws.

Using a magnetic jig allows you to remove the springs or height adjusters and avoid the problems they cause. The jointer won't be damaged by removing the parts—they simply slip out of their grooves—and they can always be put back in service later if you change your mind. If you choose to continue using the springs or adjusters, always clean them thoroughly while the knife is out of its slot and follow the setting instructions in the machine's manual.

Some jointers, mostly commercial models, have fixed-position reversible knives that make knife changing a simple job. If your machine has disposable knives, consider yourself lucky and follow the owner's manual for the proper way to set them, since the exact procedure will depend on the model you own.

FINDING TOP DEAD CENTER

Before replacing the knives, you must locate top dead center (TDC) for the cutterhead. This is the point directly above the centerline of the head. When a knife's edge is at TDC, it is at the high point of its arc and in the proper position for setting.

To find TDC on your machine, first make a locating tool as explained in the sidebar below. To use the tool, move the fence to the back edge of the table and raise the outfeed table to a point where a straightedge on the outfeed table will just clear the cutterhead with the knives rotated out of the way. When the outfeed table is set, extend the straightedge so it is over the infeed table and bring the infeed table to the same height (see the top photo on p. 40).

Making a Jig to Find Top Dead Center

YOU CAN CONSTRUCT A JIG TO HELP you find top dead center on your machine. The body of the jig is a scrap of hardwood 1½ in. wide and 4 in. long. For the jig to be accurate, the ends of the block must be square to the sides.

Drill a hole for a 1¼-in. drywall screw ⅜ in. from one corner. You should leave the screw head proud approximately ⅛ in. (see the photo at left). If the tables of the jointer are a very close fit to the cutterhead, the screw head may not fit in the gap between the cutterhead and the table. If so, substitute a finishing nail for the screw (see the photo at right).

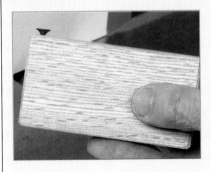

To find top dead center, begin by moving the fence all the way back and raising both tables to the same height.

You need to draw several lines to locate TDC. A piece of paper taped to the fence makes the lines easy to see.

Next, tape an index card or piece of paper to the fence to create a good surface for marking (see the bottom photo above). Place the locating tool on one of the tables with the screw or pin down in the gap between the table and the cutterhead. Slide the tool toward the head until the projecting screw comes to a stop against the cylindrical surface of the head. Holding the tool firmly against the table and cutterhead, mark the paper along the edge of the block above the pin (see the top photo on the facing page). Repeat this operation on the other table, again making a mark where the tool stops (see the bottom photo on the facing page). TDC is midway between the two marks. Locate this midpoint with a ruler and,

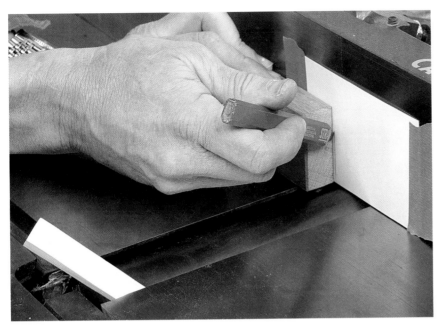

Slide the pin on the center-locating tool against the cutterhead and mark the paper where the edge of the block comes to a stop.

After using the locating tool on one side of the head, flip it over and make a second mark from the other side.

using a square, draw a vertical line on the paper to mark TDC as shown in the top photo below.

Once you have TDC marked on the fence, rotate the cutterhead until the edge of one of the knives lines up with the line marking TDC. With the knife in the TDC position, make a mark on the paper in line with the

Use a ruler to locate the midpoint between the two lines. This is top dead center.

Line up the edge of one of the knives with the TDC line and make a final mark even with the back of the knife slot. This mark is used to position the head when the knife has been removed.

back edge of the knife slot (see the bottom photo on the facing page). The purpose of this mark is to give you a reference point to position the cutterhead when the knife is removed.

CHANGING KNIVES

To change knives, start by removing one blade and its lock bar from the cutterhead (see the photo below). Most manufacturers recommend that you remove knives one at a time to avoid distorting the cutterhead.

Clean up the knife slot with solvent and a rag to remove any pitch and dirt, using steel wool or Scotch-Brite on the worst spots. Do the same with the locking bar. Next, lightly run a file back and forth a few times against the front and rear faces of the slot to remove any burrs and high spots. When the slot faces are smooth, rewipe the slot with a clean rag and solvent to remove any filings.

The face of the lock bar that bears against the knife must be flat to prevent the bar from shifting as the screws are tightened. Smooth the face

Locking the head with a pine wedge when changing knives makes it less likely that the wrench will slip.

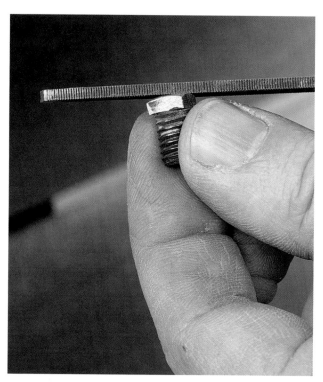

Burred and distorted lock-bolt heads are the primary cause of shifting when locking the knives in place. Clean each bolt head with a file, removing burrs and giving the bolt head a slightly rounded shape.

To prevent the knife from shifting as you tighten the lock bolts, file the face of the lock bar smooth.

of the bar using a file. If the bar is bent, straighten it out and then file the face of the bar flat (see the bottom photo on the facing page).

The locking bolt heads are almost always burred from being repeatedly tightened and loosened. These burrs are the primary reason why a blade will shift as it is tightened. You should smooth and slightly round the bolt heads, making sure the high spot on the bolt head is centered (see the top photo on the facing page). Check that the wrench is a good fit on the flats of the bolt heads, which will limit rounding and burring of the heads in the future. The wrench supplied with the machine may fit poorly, so buying a good-quality mechanic's wrench and filing the bolt heads to fit is well worth the trouble.

To install a new knife, rotate the cutterhead until the back edge of the empty slot lines up with the edge mark on the fence. Lock the head in position by tapping a small softwood wedge into the gap between the head and the infeed table. Drop the knife and lock-bar assembly into the slot and snug up the bolts until the knife is lightly locked in place (see the photo below). Now evenly loosen the bolts just until the knife can move freely.

Place the magnetic knife holders on the outfeed table, one near each end of the knife, being careful not to block wrench access to the locking bolts (see the photo on p. 46). The knife should be pulled up against the underside of the magnets. To check that the knife is free to move, push it down into the slot with a stick against the cutting edge and make sure that it springs back against the magnet as soon as you stop pushing.

(see the photo on p. 46)

Reducing Bolt Friction

Make sure the bolts turn easily when they are threaded into the bar. If not, clean out the threads with a tap. Apply a little wax to the threads and bolt heads to reduce friction as they are tightened. All other surfaces of the bars, knives, and slots should be left dry or with only a very slight oil film.

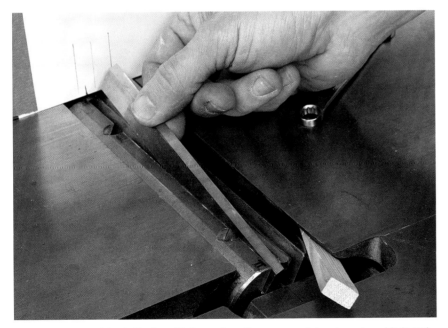

Line up the back of the knife slot with its mark on the paper and drop in a sharp blade and the lock-bar assembly.

Place the shopmade magnetic jigs on the jointer's outfeed table so the front edges of the jigs are slightly past the edge of the knife.

If you have the machine's manual, it will give you the recommended blade height. But if you lack this information, use either of the methods outlined below and you will probably be close to the proper height. No matter how you set the blade height, always rotate the head by hand after installing new blades to be absolutely sure that nothing interferes with the cutterhead.

To set the blade height when you don't have factory specifications, lower the outfeed table, with the magnetic knife holders still in place, until the back edge of the bevel on the knife drops below the outside surface of the cutterhead, then raise the table until the bevel and about $\frac{1}{32}$ in. of the back of the knife projects above the head. On most machines, this is the proper position for the knife (see the top photo on the facing page).

If the cutterhead on your machine has the back edge of the slot machined away so that you cannot judge the proper knife setting, set the outfeed table with a straightedge and feeler gauge about 0.015 in. above the cylindrical surface of the head (see the bottom photo on the facing page). Once the outfeed table is properly adjusted, don't move it until you have replaced all of the knives; this will guarantee that they are all cutting in the same arc. If the height adjustment on the table has a lock, use it to hold the setting.

Once you have the knife at the correct height, gently snug up the lock bolts. To reduce the chance of the blade shifting, go back and forth between all the bolts, tightening them progressively in three or four stages. If the blade does shift, try tightening the bolts in a different sequence or

On most jointers, when the knife height is set correctly, a narrow strip of the knife back will be visible above the back edge of the slot.

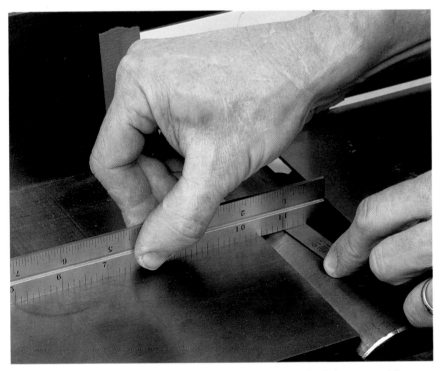

A second way to set knife height is to raise the outfeed table until a feeler gauge of the proper thickness fits between the cutterhead and a straightedge laid on the table.

tightening a couple of bolts firmly before the others. One bolt is often the worst offender, so leaving it to last will prevent it from causing mischief. After you have the first knife installed, use the same procedure to replace the other two.

Testing the Machine

It's time to test the machine. Double-check that you have tightened all the knife-lock screws, reattach the guards, and plug in the jointer. Before turning on the machine, always remove all tools from the table; vibration can cause tools to walk across the table, possibly into the cutterhead.

To set the outfeed table height, choose a test board about 2 ft. long. Lower the outfeed table to slightly below where it was set for installing the knives, then set the infeed table for a 1/16-in. cut. Start the jointer and take a pass on the edge of the board. Because the outfeed table is slightly low, the board will have an obvious snipe in its trailing end (see the top photo on the facing page). Raise the outfeed table in small steps and take additional passes until the snipe just disappears; this should be the proper height for the outfeed table.

To test the jointer for making straight edges, carefully pick two clear pine boards about 4 ft. long and from 4 in. to 6 in. wide. These boards should have no twist, and you should joint their wide face flat if necessary. On each board, the best face should run against the fence. Mark the opposite face with a V as shown in the bottom photo on the facing page, then pass both boards through the machine with the V facing out and pointing down toward the table.

Start with the board firmly on the infeed table. As soon as 8 in. to 9 in. of the board are over the outfeed table, apply most of the down pressure there with your left hand positioned just past the cutterhead. Keep your left hand in place and do all the feeding with your right hand, keeping pressure down and against the fence with your left hand as the board passes over the cutter.

Next, place the two boards edge to edge with the V's facing out and toward each other (see the top photo on p. 50). The fit should be very good. If the boards are high in the center, slightly lower the outfeed table, but if the boards are low in the center, raise the outfeed table. Be sure to adjust the table in very small increments, checking after each adjustment. Once you get an airtight joint, the table shouldn't need to be readjusted until the next time you change the blades. If the outfeed table has a lock screw for the gibs, you should tighten it to hold the setting.

To fine-tune the height of the outfeed table, start by taking test cuts with the table deliberately low. The setting will create an obvious snipe in the board's trailing edge.

Mark each of a pair of test boards with a V to indicate which side will face out and which edge will be against the table when making a test cut.

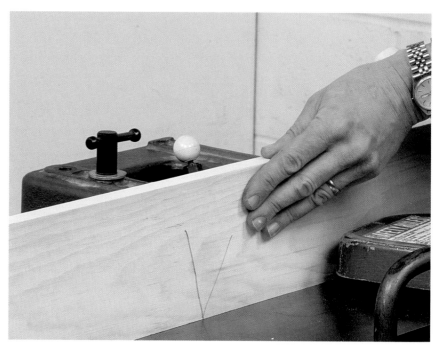

With the V properly oriented, joint the edges of both boards, taking care to keep the cut edge firmly against the outfeed table through the entire cut.

Stand the test boards together edge to edge with the V's pointing at each other. If the joint isn't perfect, a slight adjustment of the outfeed table will correct the fit.

Placing a straightedge across the joint will show whether the fence is square to the table. Readjust the fence stops if needed, and run the test boards again to check the setting.

Once you are cutting straight, place a straightedge across the joint. If the fence isn't quite square to the table, there will be a slight bow or crown. Fine-tuning the fence's vertical stop will eliminate the problem (see the photo above).

The last step is to set the depth of cut scale for the infeed table. Place a short straightedge on the outfeed table and extend it over the infeed table, then raise the infeed table until it just touches the straightedge. Now position the pointer for the depth scale to read zero and the jointer is ready to use.

The Table Saw

A table saw is at the heart of most woodworking shops. It is a simple, sturdy machine used mainly for ripping and crosscutting, but it also can be used to create a variety of woodworking joints, moldings, and raised panels.

In cabinetmaking, a fraction of a degree or a few hundredths of an inch can make the difference between first-rate workmanship and something that seems sloppy or poorly made. A well-tuned table saw allows you to work accurately to very close tolerances and to do it confidently. Even if the saw is used only for rough work, it should be reasonably well tuned. A badly adjusted saw is frustrating and, worse, dangerous to use.

Tuning up a table saw is a straightforward job that usually can be completed in an afternoon. Your reward will be a saw that works smoothly and cuts stock accurately, and your sawblades will last longer because they won't overheat.

Table-Saw Types

Table saws fall into three general categories: cabinet saws, contractor's saws, and benchtop machines. Except for the location of the motor and general heft of the machine, cabinet saws and contractor's saws are very similar in design. Modern benchtop saws, which use compact motors and aluminum castings, are distinctly different.

Cabinet saws are the heaviest and most powerful of the group. They have an enclosed steel base, which gives them their name, a cast-iron table, and motors of 2 hp to 5 hp. Smooth running and powerful, cabinet

A cabinet saw has an enclosed base and a powerful motor. Its heavy castings contribute to smooth, vibration-free cuts.

A contractor's saw is designed so it can be moved to a job site, but it is lighter and less powerful than a cabinet saw and may show more vibration.

Benchtop table saws offer portable yet sophisticated designs that make them useful for cabinetmaking.

Cabinet-saw components, including its massive motor and trunnion assembly, are attached to the base, not the top. The design simplifies some tune-up adjustments while offering better access to working parts for major repairs.

saws have been around since the 1930s and are essential equipment in most professional shops. Typically supplied with cast-iron extensions (also called wings), they are often supplemented with laminate-covered extension tables, which make the saws more convenient and safer to use.

The Table Saw

A cabinet saw is heavy and powerful, and the enclosed base aids dust collection. The rip fence on this saw locks on both the front and rear rail; some fences lock only at the front.

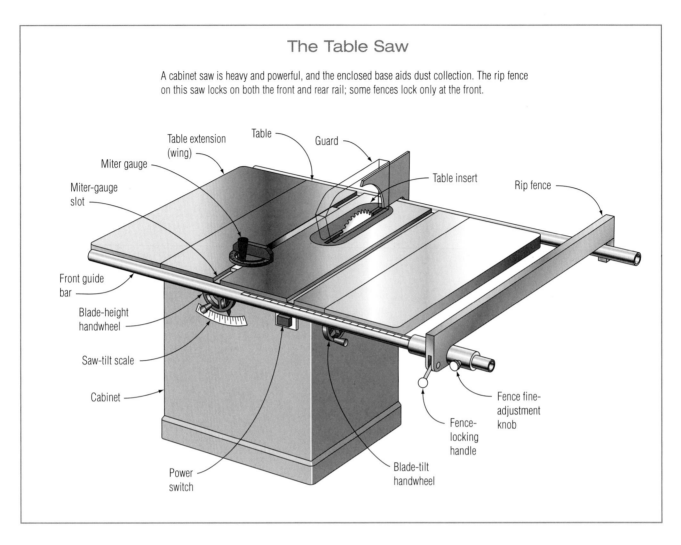

Table extension (wing)

Table

Guard

Miter gauge

Miter-gauge slot

Table insert

Rip fence

Front guide bar

Blade-height handwheel

Saw-tilt scale

Cabinet

Fence fine-adjustment knob

Fence-locking handle

Power switch

Blade-tilt handwheel

Because the running gear is mounted on a single, large casting, cabinet saws tend to be stable and therefore very accurate. The extra stability also means saw parts are not as likely to rattle out of alignment. From a maintenance point of view, cabinet saws have an essential advantage: The motor and arbor assembly is attached to the base cabinet rather than to the top (see the bottom photo on p. 53). This makes it easy to align the miter-gauge slot with the blade. An enclosed base also improves dust collection.

Contractor's saws were designed so carpenters could take them to the job site. Weight-saving design compromises, however, reduce the saw's accuracy and increase vibration. Cast-iron components are lighter than those on cabinet saws, and they are not made from a single casting. Contractor's saws come with either cast-iron or stamped-steel table extensions, although extensions on some older saws were open cast-iron grids. Motors on contractor's saws are suspended from a bracket at the back of the machine, allowing it to be removed for transportation (see the photo at left on the facing page).

The motor on a contractor's saw pivots from the back so its weight tensions the single drive belt. Although the design makes it easy to remove the motor, it also can create additional vibration, especially when the belt is worn.

Dust collection is often excellent in newer benchtop table saws, thanks to a die-cast aluminum housing that fully encloses the blade.

Vibration is the most obvious liability with this design, and it shows up in rougher cuts. Contractor's saws also can be hard to tune because the arbor assembly flexes under the weight and torque of the motor. Both problems can be addressed, but contractor's saws are not the best choice for precision work, and their open design makes dust collection more difficult.

Modern benchtop saws, which have plastic bases, aluminum die-cast tops, and high-speed universal motors, have been around for almost 30 years, but they have only recently grown in size and sophistication to become a useful tool for cabinetmakers. The smaller motors in these machines achieve their high-horsepower output by turning at better than 10,000 rpm. The gears or cogged belt inside the housing are designed to bring this down to a more sedate 3,000 rpm to 4,000 rpm at the arbor shaft.

Benchtop saws are not as rugged as cast-iron saws, but the best of them now have full-sized miter-gauge slots and reliable rip fences. The design, finish, and operation of these machines is usually excellent. Benchtop saws tend to have adequate power, but the motors make an unpleasant, high-pitch whine. Dust control, however, is very effective because the blades are enclosed in a shroud that directs sawdust into a vacuum port (see the photo at right).

Inside a Table Saw

Trunnions on cabinet saws and contractor's saws allow the blade to pivot for making bevel cuts. Contractor's saws have a single belt connecting the rear-mounted motor and arbor pulley; cabinet saws typically have multiple belts with the motor inside the cabinet.

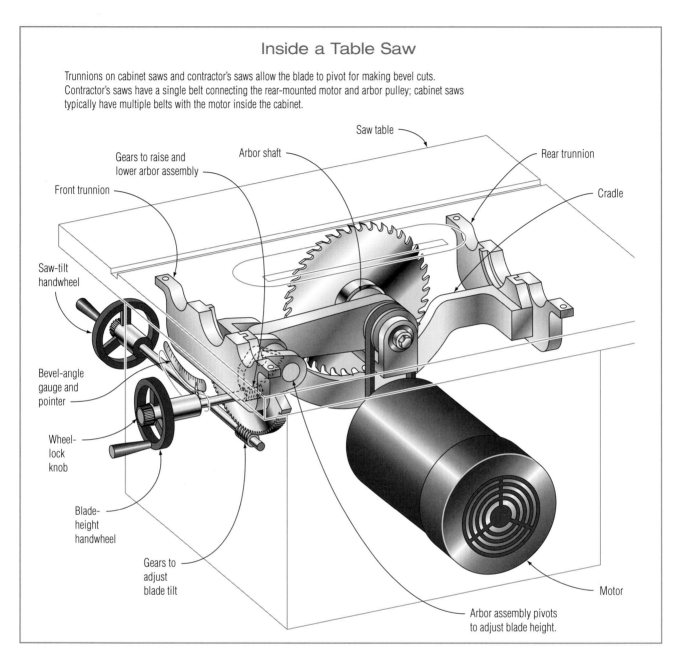

Saw table

Arbor shaft

Rear trunnion

Gears to raise and lower arbor assembly

Cradle

Front trunnion

Saw-tilt handwheel

Bevel-angle gauge and pointer

Wheel-lock knob

Blade-height handwheel

Gears to adjust blade tilt

Motor

Arbor assembly pivots to adjust blade height.

How the Saws Work

For all their versatility, table saws have relatively few controls. One handwheel raises or lowers the blade to adjust the depth of cut, and a second handwheel tilts the blade to one side to make angled cuts. The on/off switch is either mechanical or magnetic. The blade itself is mounted on an arbor and held tightly against a flange with a heavy nut and washer. On contractor's and cabinet saws, power is transmitted to the arbor via one or more V-belts and pulleys. On higher-horsepower cabinet saws, there may be two or three parallel belts to handle the increased power. Benchtop machines are direct drive, meaning that the end of the motor shaft is the arbor (a few have an intermediate gear).

To tilt the sawblade, the arbor shaft, its housing, and the motor assembly must move as a unit. On cabinet saws, the arbor and motor assembly, which can easily weigh 100 lb. or more, hangs from trunnions at the front and back of the machine. At each trunnion, curved tongue-and-groove parts support the weight of the assembly while allowing it to tilt (see the illustration on the facing page).

To adjust blade tilt in cabinet saws, contractor's saws, and some benchtop machines, a handwheel-operated worm gear drives a second gear mounted on the arbor assembly. The arbor on a cabinet saw is mounted on a swinging arm. To adjust blade height, a handwheel turns a set of gears and pivots the arm up or down.

Benchtop saws, which don't have the mass of cast iron to support, use simple hinge pins to support the motor and arbor unit under the table (see the illustration on p. 58). To keep the sawblade centered in the slot at all angles, the hinges on these saws must be kept small and tucked up very tightly under the tabletop. Benchtop saws also use a very different system to control the blade height. The entire motor and transmission unit slides in tracks that are molded into the main casting. The unit is controlled by a threaded rod and nut that are mounted along one side of the slide assembly. The rod is rotated by a handwheel on the front of the saw.

Almost all new table saws, regardless of type, use 10-in. blades. Older machines sometimes have 8-in. or 9-in. blades. The craftsmanship on some of these earlier machines, by the way, is superb, and they can represent a real bargain if you find one in good condition and you can live with the reduced depth of cut.

Tuning Up the Table Saw

A complete tune-up involves a number of steps. It makes sense to clean and lubricate the saw's working parts first, before checking and adjusting the saw's components and settings. I take the work in the following order:

- Pulleys, belts, and bearings
- Table extensions, table insert, and splitter
- Alignment of blade and miter-gauge slot
- Rip fence
- Blade-tilt settings
- Miter-gauge stops

Some of these steps—such as making sure the miter-gauge slot and the rip fence are parallel to the sawblade—take some tinkering, yet they are essential. Also essential is checking blade-tilt settings and miter-gauge

Modern Benchtop Saws

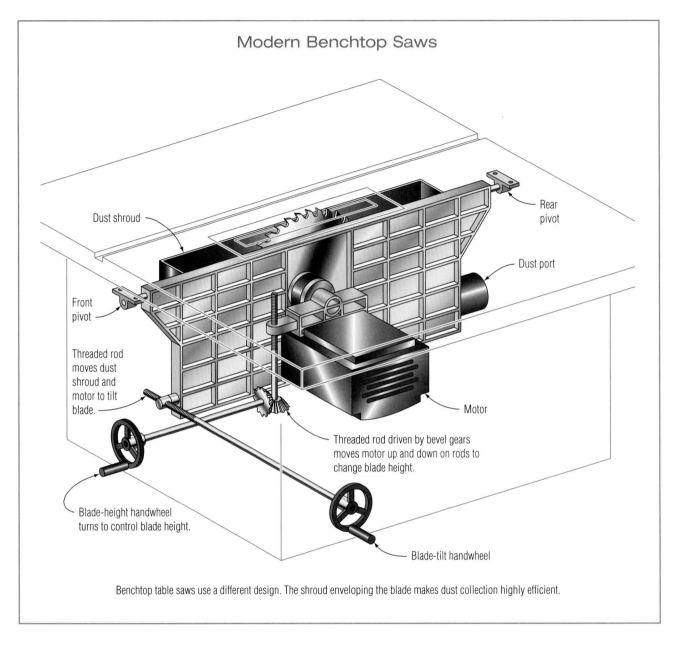

Dust shroud

Rear pivot

Dust port

Front pivot

Threaded rod moves dust shroud and motor to tilt blade.

Threaded rod driven by bevel gears moves motor up and down on rods to change blade height.

Motor

Blade-height handwheel turns to control blade height.

Blade-tilt handwheel

Benchtop table saws use a different design. The shroud enveloping the blade makes dust collection highly efficient.

stops for square and 45 degrees. If you get these elements right, your saw will consistently cut stock square and straight.

It may not be glamorous to start by cleaning and lubricating the saw's moving parts and inspecting belts, pulleys, and bearings, but a clean, well-oiled machine is easier to work on, and it will hold its settings longer and more accurately. Just as important is to double-check the settings that were made when the saw was put together: the extension wings, the table insert, and the splitter.

Except for aligning the miter-gauge slot with the blade, the tune-up and basic setup procedures can be done separately anytime you think something has slipped out of line. Adjusting a saw's miter-gauge slot alignment, however, will throw off the rip fence and miter-gauge settings, so always begin a full tune-up by checking and aligning this key setting before moving on to the fence and miter-gauge adjustments.

There are hundreds of models of table saws, some with unique design features. The maintenance procedures I describe here are for the most common saws, but you may have to do some sleuthing on your own machine to find adjustment screws and lock bolts. Before beginning, you might find it helpful to spend some time with a flashlight under your machine to identify its components and understand how they function.

Choosing the Right Blade

In addition to a good tune-up, choosing the correct blade is a key to getting the best performance from your table saw. You can do a lot of agonizing over which sawblade to purchase: a rip, a crosscut, or a combination. Are 60 teeth better than 20? Should the tooth pattern be an alternate-top bevel or triple chip? And is any blade really worth $100? (For help in choosing a blade, see the sidebar on p. 60). Actually, it's all pretty simple, including the question of price. With sawblades, you really do get what you pay for.

Almost all of the better-quality blades made today have carbide teeth (see the sidebar on p. 61). Although all-steel blades have been nudged aside, they did have some advantages. They can produce a superb cut because the teeth can be sharpened in profiles that are impossible with brittle carbide. Steel teeth also can be sharpened easily right in the shop or at low cost by a local sharpening shop. But the increased use of highly abrasive man-made materials, such as MDF, as well as the falling cost of carbide tooling have made carbide-tipped blades the standard.

Initially there were only two types of sawblades for wood—rip blades for cutting with the grain and crosscut blades for going across the grain. You would have to change from a rip blade to a crosscut blade depending on which type of cut you were making. Blade manufacturers, seeing a big market, developed the combination blade, which can both rip and crosscut.

Ragged or burned cuts, kickback, and difficulty feeding the board can be caused by a warped, dirty, or dull blade or by using the wrong blade for the job. A quick test is to switch to a different blade: If the problem clears up completely, it's obvious what the problem was. Little or no improvement suggests alignment problems.

Flat-Top Grind

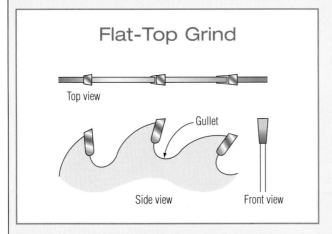

Alternate-Top Bevel

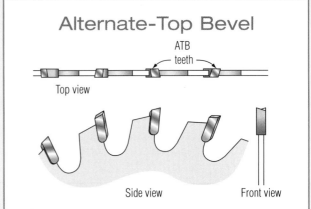

YOU WILL GET THE BEST RESULTS FROM YOUR TABLE SAW by choosing the correct blade for the job you're doing. The four types are rip blades, crosscut blades, combination blades, and triple-chip blades.

Rip blades, which are designed to cut with the grain, have a flat-top tooth profile, called a flat-top grind (FTG). Each tooth takes a long, planelike shaving, while the blade's deep gullets clear debris quickly. There are typically no more than 24 teeth on a 10-in. blade. Rip blades are especially good on stock thicker than 1¼ in. and on tough wood such as oak and tropical hardwoods. Rip blades are a necessity for cutting green wood and some softwoods because the stringy shavings clog up the smaller gullets of combination blades.

A crosscut blade has a different tooth shape. The profile comes to a sharp point that cleanly slices the wood fiber. Teeth are beveled across their width. Since the blade needs to cut along both edges, the bevels reverse their slope from tooth to tooth, hence

the name alternate-top bevel (ATB). Sixty teeth on a 10-in. blade is common. Crosscut blades do well at cutting MDF and plywood but may cause some chipping of veneers and melamine facings.

A combination blade is exactly that, a mix of alternate-top bevel teeth and flat-topped teeth called rakers. The typical ratio is four ATB teeth followed by a flat-topped raker with a deeper gullet. The tooth count falls in the mid range, 30 to 40 teeth in a 10-in. blade. Some combination blades have ATB teeth only, without the rakers. They rip somewhat more slowly than the blades with rakers and probably crosscut a bit better. A combination blade works well with both solid wood and man-made materials, but it may bog down when used for heavy ripping and may cause some chipout on veneers.

Triple-chip blades are designed for very hard materials that chip easily, such as melamine. Teeth beveled at each corner alternate with rakers, which may be either flat topped or in an alternate-top bevel design.

Alternate-Top Bevel and Raker

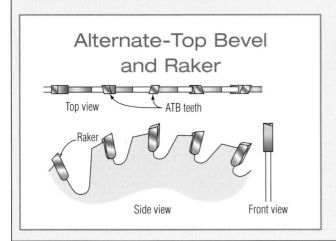

Triple-Chip Grind

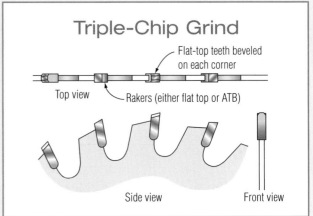

CARBIDE IS A SIMPLE CHEMICAL compound consisting of a metal, typically tungsten, and carbon. Pure carbide comes in the form of microscopic granules. Grains of carbide are cemented together with a binder, such as cobalt, to form saw teeth.

A carbide cutting edge is extremely hard, so it doesn't wear away very quickly. Heat and chemicals in wood resin and adhesives, however, attack the binder. Once the binder fails, the carbide grains break free and the cutting edge loses its shape. Carbide cutters also fail by fracturing. Sawing through especially hard knots, common in some tropical woods, or striking embedded gravel can shatter a tooth. Most blades, in fact, have a few chipped teeth after they've been used for a while.

The amount and type of binder and the size of the carbide grains affect both chemical resistance and toughness. Unfortunately the more chemical-resistant combinations tend to be more brittle. The best, most highly engineered carbides are the most expensive. This is part of what you pay for in a top-of-the-line blade.

A mounting block for a dial indicator (top) and a plate made from plywood are essential but easy-to-make tools for aligning the miter-gauge slot and the rip fence to the blade.

Assembling the Right Tools

For checking both the alignment of the miter-gauge slot and the rip fence and for checking arbor runout, you will need a dial indicator with a shop-made mounting block and a rectangular piece of Baltic birch plywood. You may need some extension tips for your dial indicator, but they don't have to be expensive. You also will need an accurate square, either a machinist's or combination style, and an assortment of wrenches, screwdrivers, and hex wrenches.

The mounting block shown in the photo above is a piece of straight-grained pine. It should be 1 in. to 1½ in. thick and cut to the shape shown

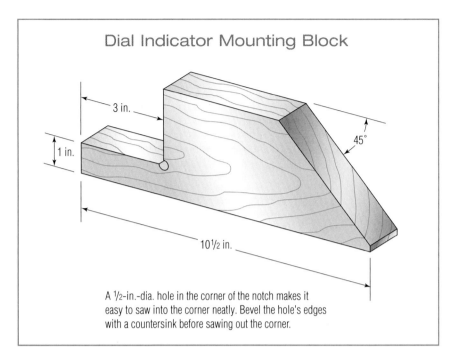

Dial Indicator Mounting Block

3 in.

1 in.

45°

10½ in.

A ½-in.-dia. hole in the corner of the notch makes it easy to saw into the corner neatly. Bevel the hole's edges with a countersink before sawing out the corner.

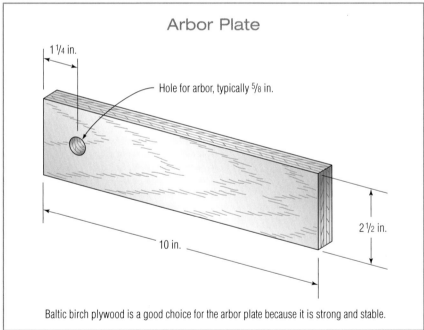

Arbor Plate

1¼ in.

Hole for arbor, typically ⅝ in.

10 in.

2½ in.

Baltic birch plywood is a good choice for the arbor plate because it is strong and stable.

in the top illustration above, but it's not critical that you use these exact dimensions. Beveling all the edges will keep them from getting nicked. The dial indicator may be mounted in several positions on the block, depending on the test you are performing (see the photos on the facing page). It's better not to attach the indicator until you are setting up each test.

When clamped to the block this way, the dial indicator is used to check the alignment of the miter-gauge slot to the blade.

Attached to the sloped end of the block, the indicator can be used to check for any wobble on the arbor flange.

Mounted in place of the blade, the plywood plate (see the top illustration on the facing page) will be used to check the alignment of the arbor with the miter slot and the rip fence. Baltic birch is cabinet-grade material with smooth, hardwood veneer on both faces. I use a piece ½ in. thick. This kind of plywood has more laminations than ordinary plywood, making it strong and stable. The hole should be sized for your arbor, which is ⅝ in. on most modern saws. Lightly sand both faces of the plate, but don't apply a finish because the wood needs to show a pencil mark. Draw a centerline down the better face of the plate.

You might wonder whether a plywood plate is accurate enough to be used as a reference for checking the accuracy of machine parts. After all, a dial indicator can detect very small amounts of wobble, and several companies sell $50 precision steel disks for the same purpose. There *would* be a problem if the indicator were used to measure more than one place on the plate; the slight variations in thickness would make accurate measurement impossible. My method skirts this problem because all measurements are made against the same spot.

Moreover, the rectangular shape of the arbor plate extends well beyond the blade opening so front and rear measurements are taken at locations much farther apart than they would with a steel disk. As a result, the effects of any misalignment are larger and easier to measure. I'll explain how to use the plate in "Aligning the Blade and Miter-Gauge Slot" on p. 73.

Cleaning and Lubricating

Like all machinery, a table saw benefits from regular cleaning and lubrication. In a busy commercial shop, this could be as often as every few weeks. In a home shop, once or twice a year is adequate, perhaps in the fall after you've taken a summer break from woodworking or just before starting a big project. Start by donning a dust mask and removing the bulk of the sawdust inside the machine with a shop vacuum. Use a hand-held brush to remove any dust in the corners you can't reach with the vacuum nozzle.

Once you've gotten as much debris as you can with the vacuum, use an air compressor to clear away the last of the dust from the hidden recesses of the machine. If you don't have one, switch the vacuum hose to the exhaust port and use the nozzle as a blower.

Lubricating woodworking machinery in general is problematic and table saws are the worst of the lot. On the trunnions and adjusting gears, lubricants such as lithium or bearing grease instantly load up with sawdust. The grease is pushed aside in useless, hard-to-remove clumps. Oil, of course, is an even worse dust magnet. I confess that I've resorted to a quick shot of penetrating oil to relieve binding or squeaking, but the effect doesn't last long.

The best lubricant I have come up with for exposed bearing surfaces on a saw is a white stick product in a push-up tube that I found at a local hardware store. It eliminates friction, but it doesn't attract dust as readily as grease. It looks and feels like slightly soft bar soap.

Before you apply new lubricant, try to remove most of the old, especially the dense dust and grease mixture that often encrusts the works of older machines. Once you've cleaned off the worst of the old grease, apply the new sparingly. It doesn't take much. The working layer of lubricant is only a few molecules thick, so any surplus will just attract dust and dirt.

Use the stick lube on the worm and gear and trunnions as you run the height and tilt mechanisms from stop to stop. Apply a drop of light machine oil to the control shafts where they pass through the castings and support bushings. Any moving part that is impossible to reach directly should get a shot of a spray lube through one of those easily lost little tubes that mount on the spray nozzle.

The Future Is Plastics

Bushings, gears, and mating surfaces in many of the new bench-top machines are made from plastic. These parts don't need lubrication and can actually be harmed by the solvents in conventional greases and oils. On these surfaces, use silicone-based greases or sprays sparingly if at all.

Checking the Saw's Drive Line

It is unusual to have problems with the bearings in the motor or the arbor shaft. But problems with belts and pulleys are common, especially on contractor's saws. Depending on how the saw is constructed, replacing a bad belt or even tightening a loose pulley can require a considerable amount of work. In some cases, the saw must be disassembled. If repairs are needed on a cabinet saw, it is often easier in the long run to remove the table to gain complete access to the arbor and motor assembly. Contractor's and bench-top saws can be flipped over for servicing.

To avoid injury when working on a table saw (or any other machine, for that matter), make sure the tool is unplugged whenever you will be

Bearings in good condition will allow the arbor to be rotated smoothly and silently. When the arbor catches or when a clicking noise is audible, faulty bearings are typically to blame.

near the blade or other moving parts. Before unplugging the saw, however, run it both with and without the sawblade and listen and feel for unusual noise and vibration. A cabinet saw should make only a low steady hum and almost no vibration; a contractor's saw will be noisier on average but there should be only minor vibration.

The motors of benchtop saws run at high speed and produce a high-pitched whine. Listen for intermittent buzzes and vibration that come and go or a steady, fast rattle. These are a sign that bearings may be worn out. The drive on these saws is an enclosed cogged belt or gears that rarely have problems, but a cogged belt or gear drive that is failing will produce sudden jumps and skips when sawing.

With the power disconnected and the blade removed, rotate the arbor by hand (see the photo above). The shaft should rotate smoothly with little noise. If you hear a grinding or clicking sound or if the shaft sticks

The drive belt should be smooth and flexible with no signs of cracking.

abruptly, a ball bearing has probably reached the end of its life. To isolate the problem, disconnect or slack off the belts and rotate the motor and arbor separately. This is easy to do on a contractor's saw, but it's somewhat difficult on a cabinet saw.

A second test to run while rotating the arbor is to pull up sharply on the shaft as you rotate it. There should be no motion or clicking that would indicate worn bearings. As best you can on contractor's and cabinet saws, repeat this test on the pulleys of the motor and arbor shaft. Any motion here is probably due to a loose pulley. If you can reach the set screws on the pulleys with a wrench, it is worth checking that they are tight.

On a contractor's saw or cabinet saw, the arbor should have little or no side-to-side play, that is, play along its axis. If there is, it can probably be eliminated by adjusting the pulley or a large nut threaded on the shaft behind the pulley. A benchtop saw, if it is gear driven, may normally have some end play in the arbor shaft; under a load, this goes away.

A stiff or cracked belt will be felt as the arbor being alternately easier and then harder to turn, typically only once in several revolutions of the arbor as a bad spot on the belt reaches a pulley. A badly deteriorated belt will create severe vibration when the saw is running. Visually examine the sides of the belt as you move it. Look for cracks or patches of missing rubber and shiny spots that indicate the belt is hardening and slipping (see the photo above).

On a cabinet saw, press on the back of the belt midway between the pulleys; it should flex only slightly. If there's more slack, adjust the belt tension by adjusting the motor mount. On a contractor's saw, the belt is tensioned by the weight and position of the motor. Shifting the motor mounts to better align the pulleys and increase the belt tension often makes these saws run more smoothly.

Contractor's saws often suffer from a combination of poor-quality or worn pulleys and stiff belts. The end result is that the motor, which is cantilevered off the back of the saw, bounces and vibrates. This affects the quality of the cut and quickly throws the saw's arbor out of alignment. Inline Industries markets a kit with two machined steel pulleys and a Power Flex Plus belt that can improve performance considerably.

Checking the Saw's Original Settings

When a saw is brand new, its various components must be assembled into a working tool. This process includes attaching the table extensions, adjusting the table insert, and making sure the splitter is aligned correctly. Over time, these parts can drift out of alignment, so a full tune-up should include a look at all these parts and, if necessary, some readjustments. The insert and splitter are easy to adjust, but getting the wings to line up flat with the saw's main table may require some shimming.

LEVELING A TABLE SAW'S WINGS

Leveling a table saw's wings can be difficult because mating surfaces do not always line up perfectly. But taking it slowly and methodically and using a few shims make this a relatively painless procedure. The process I describe here of leveling the wings with shims is intended for cast-iron extensions. Getting stamped-steel extensions—like those on some contractor's saws—in the same plane as the table can be accomplished by bending the wings once they are bolted to the saw. Stamped wings are too flexible to stay in line, however. They also need to have braces added so they remain rigid under a load.

Attaching a cast-iron wing is an awkward job because the casting is too heavy to support with just one hand while you get the bolt started. Adding a knee to the process makes it much easier. Crouch down next to the saw with one of your knees directly below the wing's center bolt hole in the side of the saw's table, then stand the wing vertically, on its end, on top of your knee. The center hole in the wing casting should line up with the hole in the table. Sometimes placing a block or two of scrap under your foot, to raise your knee, will make it easier to get the holes in line. Now with one hand and your knee supporting the wing, you still have one hand free for the bolt.

To level the wing, you should attach it to the table with all three bolts. Front and rear bolts should be snug; the middle one should be loose. Tap the table with a nonmarring hammer until the front edge of the wing is flush with the front of the table (see the photo on p. 68), then tighten the front bolt to three-quarters tight to hold it in place.

| On the Level |

Before adjusting the wings, take a look at how the saw sits on the shop floor. The saw's base or stand should be equally supported at all four corners. If the saw is supported unevenly, stress will distort the machine all the way up to the table's surface. Use shims or add screw adjusters to the bottom of the legs to even up the load.

With the attachment bolts snug but not tight, line up the front edge of the wing and the top with a few light taps using a nonmarring hammer.

Next, move the back of the table up or down until it is flush over the rear bolt, and tighten the bolt to hold the table in position. Go to the front of the table and readjust it, if needed, so it's flush, both over the bolt and along the front edge. Go back and forth between the front and rear, making these adjustments until the top surface of the wing is flush at both ends and the front edge is flush, then fully tighten all three bolts.

Place a short straightedge over each bolt and note whether the joint between the table and the wing is flat, crowned, or dished. Typically, there is a crown because the outside edge of the wing sags slightly. If the wing is not on the same plane as the table, it should be shimmed. The aluminum from a soda can makes good shim stock, which you can cut with scissors. Cut the shims ¾ in. wide and 1½ in. long, and fold a ¼-in. tab along the long edge of each shim.

Now loosen all three bolts slightly, but not enough to allow the wing to lose its alignment to the table. Fully loosen one bolt and insert the ¼-in. tab of the shim between the table and wing. The shim should be centered on the bolt (see the top photo on the facing page). You may have to flex the table with hand pressure to get the shim to slip in place. Insert the shim from below if the wing sags or from above if the wing rises. Retighten the bolt three-quarters tight.

Insert the shims at the remaining two bolts, checking that the wing and table are still flush at the joint, then fully tighten the bolts. Check to see if the wing and table are now in the same plane. If they are not

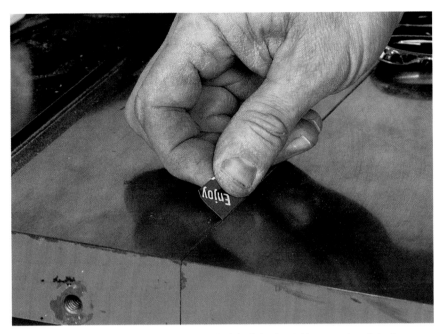

Shims cut from an aluminum soda can are ideal for bringing table-saw wings into the same plane as the main top.

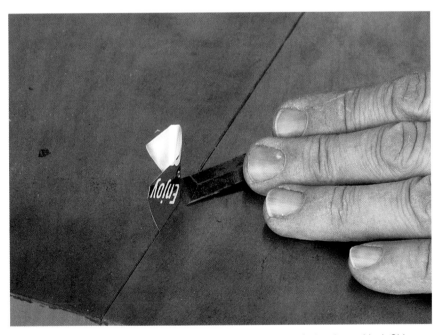

Shims installed from the top should be cut flush with the top using a sharp chisel. Shims installed from the bottom are out of the way and may be left exposed.

in line, increase or decrease the thickness of the shims. Once the surfaces are flush, trim any visible shims using a chisel (see the bottom photo above).

Turn the set screw to raise or lower the insert until all four corners are slightly below the level of the table.

Sliding a small hardwood block over the edge of the insert is a good test: If the block catches, the insert is too high.

Tighten Up

Leveling screws in a table saw's throat plate may not have locknuts. A small plug of plastic may be inserted in the threads to keep the screws from backing out, or they may be held in place with a special adhesive. If an adjustment screw won't stay put because the insert or adhesive has failed, try using a spot of rubber cement on the threads.

ADJUSTING THE TABLE INSERT

When the table insert isn't perfectly flush with or slightly below the top of the table, it can cause a board to bind during a cut. Adjusting an insert, however, is straightforward. It helps to have a small, hardwood block on hand for checking the fit. Cut one roughly 1 in. by 2 in. by 3 in. and put it on the table close to one of the adjusting screws that you'll find in (or under) the insert. Slide the leading edge of the block back and forth over the joint between the plate and the tabletop (see the photo at right above). If the insert is too high, you will feel the block catch on its edge. Turn the screw to lower the plate's height until the insert is just below the edge of the opening. Repeat this procedure all the way around the plate at each adjustment screw (see the photo at left above).

Once the plate is below the table height, place the block on the table near one of the screws at the infeed end of the plate. Slide the block back and forth while adjusting the screw to raise the plate until it just catches the block. Now back off the screw, just slightly, until it no longer catches; this is the proper height of the insert. After making the same adjustment at each screw location, the insert height should be correct.

ADJUSTING THE SPLITTER

Some boards are contorted by so much internal tension that they try to close around the blade during a cut. This dangerous condition, which can lead to kickback, is what a splitter is designed to prevent. The splitter, in some ways, is misnamed; it was never intended to force apart or separate the two halves of the stock. Its true purpose is to stop kickback by preventing the stock on the outfeed side of the table from making contact with the teeth at the back of the sawblade.

The splitter should be slightly thinner than the blade and directly in line with it. When ripping a board in two, neither piece should touch the splitter. Every saw maker has a slightly different method of mounting and adjusting the splitter, and you'll have to work out or look up the exact method of setting up the one on your machine. Unfortunately, many splitters are so poorly designed and flimsy that they can't be made to line up properly and hold a setting.

To check the splitter's alignment, cut two 1x2 boards so they are long enough to reach from the front of the blade to a point 6 in. beyond the splitter. Joint one face and one edge flat and square and mark the good faces. Then unplug the saw and mount a fine-tooth blade on the arbor, but don't replace the table insert.

Check that the blade is in the vertical position and close to full height. Place the boards on either side of the blade, making sure they bear on a tooth at both the front and the rear of the blade (see the photo at left below). The splitter should be in the middle of this gap, not touching either board. If not, you will have to adjust it.

With one board removed, check whether the side of the splitter is parallel to the edge of the board. If these two surfaces are not in line, the vertical alignment of the splitter will need adjusting (see the photo at right below). If the splitter's mounting bracket doesn't have an adjustment for the vertical alignment, you may need to bend or twist the splitter to cor-

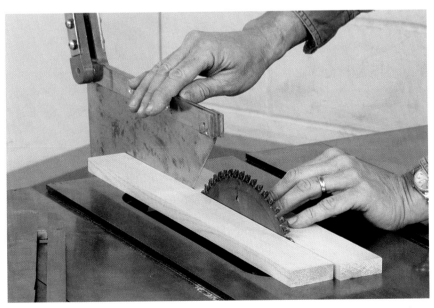

A pair of jointed boards will quickly show whether the splitter is correctly aligned with the blade. The splitter, slightly thinner than the blade, should not touch either board.

Adjust the splitter so it is parallel to the board's vertical edge. Some splitters will have to be gently bent into position.

Adjust the splitter so it is parallel to the board's vertical edge. Some splitters will have to be gently bent into position.

rect its tilt and get it parallel to the plane of the blade (see the photo at left). Once the vertical alignment is right, adjust the nuts or shims that center the splitter to get it midway between the two boards.

Checking for Arbor-Flange Runout

The blade is mounted on the arbor and clamped against the flange by a nut and washer. If the face of the flange isn't flat, the blade will wobble as it rotates. A wobbling blade will produce a rough surface on the workpiece and make the saw work harder. It won't affect the squareness of the cut or cause burning of the cut surfaces; these problems are due to a misaligned rip fence or a badly warped blade. An out-of-true flange is rare and not something you'd have to check in an ordinary tune-up, but it is worth checking at least once, just to be sure. The amount that the flange is out of true is called runout.

Dust, resin, and burrs on the flange also can cause the blade to wobble, so before you test for runout, it's a good idea to clean the face of the flange using solvent on a rag, then lightly run a file over it to knock down any rough spots. Once that's done, you can measure runout. Here's how I do it:

Unplug the saw and remove the blade, then tilt the saw to the 45-degree stop and raise the arbor to its highest position. Mount the dial indicator on the sloped face of the block. Place the miter gauge in the slot that the flange is facing, and position the block against the face of the miter gauge so that the tip of the gauge can bear against the face of the flange near its perimeter (see the photo at left on the facing page). On some saws it may be necessary to place an accessory tip on the indicator to reach the flange.

Slide the block toward the flange until the indicator has made a couple of revolutions and the hand stops at the 12 o'clock position. Clamp the block to the miter gauge, then zero the indicator while holding the block with one hand. Holding the block steady, rotate the arbor, preferably by pulling on the drive belt or the pulley to avoid bumping the stem of the indicator. The difference between the widest swings of the hand is the runout of the arbor (see the photo at right on the facing page).

On a well-made saw that hasn't been abused, runout is typically no more than 0.003 in. at the rim of the flange. If you are getting higher readings, it's worth trying to correct it, especially if you aren't getting good sawn surfaces when using a high-quality blade. Correcting excessive runout by replacing the flange and possibly the arbor shaft is an option, but it requires a tear-down of the machine and a substantial outlay for parts. An alternative is to have the original arbor and flange machined true, but this still requires disassembling the saw.

Runout also can be corrected by filing down the high spot on the flange. Use a new file with the teeth ground off the narrow edges to pre-

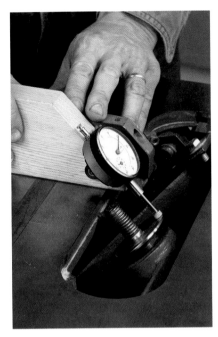

Tilting the blade to 45 degrees makes it easier to check runout on the arbor flange.

After clamping the mounting block and dial indicator to the miter gauge, rotate the arbor slowly by hand. The difference between the widest swings of the hand is the runout of the arbor flange, which should be no more than 0.005 in.

First Things First

Adjusting a saw's miter-gauge slot alignment will throw off the rip fence and miter-gauge settings. A full tune-up should start with aligning the miter-gauge slot with the blade, followed by checks of the fence and miter gauge.

vent nicking the arbor shaft. As you file, check your progress with the dial indicator. It is tedious work but worth the effort.

Aligning the Blade and Miter-Gauge Slot

When the miter-gauge slot is not parallel to the blade—and you can't assume that it is—you'll get rough, burned, and out-of-square crosscuts. No amount of fiddling with the miter gauge seems to correct it. Because this problem is somewhat hard to check and rarely discussed in owner's manuals, many woodworkers are unaware of its importance or that it can be adjusted.

When the miter-gauge slot is not aligned with the blade, stock runs into the body of the blade as the cut progresses (see the illustration on p. 74). This causes three things to happen. First, the friction generates enough heat to burn the leading edge of the board and overheat the blade. Second, the stock is forced sideways, leading to a slight curve in the sawn edge. Finally, the board strikes the teeth at the back of the blade, risking kickback as the upward motion of the saw tries to lift the board off the table.

Adjusting the alignment on cabinet saws and most benchtop saws is fairly easy. On these machines, it is worth the trouble to adjust the align-

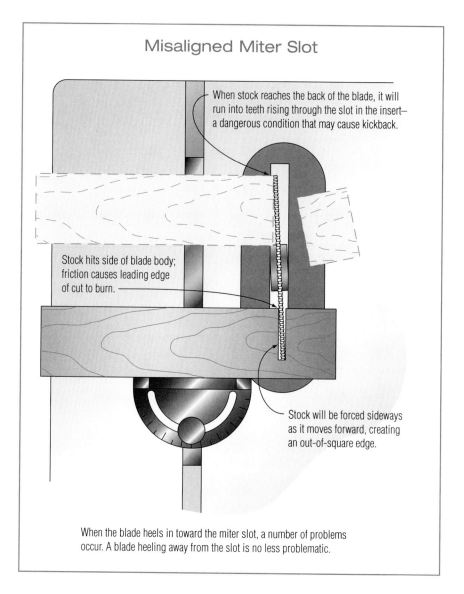

Misaligned Miter Slot

When stock reaches the back of the blade, it will run into teeth rising through the slot in the insert— a dangerous condition that may cause kickback.

Stock hits side of blade body; friction causes leading edge of cut to burn.

Stock will be forced sideways as it moves forward, creating an out-of-square edge.

When the blade heels in toward the miter slot, a number of problems occur. A blade heeling away from the slot is no less problematic.

ment so there is only a few thousandths of an inch difference between the front and rear measurements. In fact, it's not that hard to get a near zero difference between the front and rear readings. Contractor's saws are harder to adjust and tend to go out of alignment easily, but you should try to get the difference between the front and rear readings within 0.005 in. if possible. On contractor's saws, always take your measurements with the motor and belt attached because the pull of the belt shifts the arbor's position significantly.

Depending on how carefully your saw was made, the miter-gauge slots may or may not be exactly parallel with each other. If they aren't, cuts made with the miter gauge in the slot you adjusted parallel to the blade will be good, while cuts made with the other slot may still have problems. Other than replacing the top, there is no solution to this problem. Take your measurements from the slot you use most of the time.

CHECKING THE SLOT ALIGNMENT

First, unplug the saw, remove the blade, and place the miter gauge in the slot you regularly use for making crosscuts. Raise the blade to just below its highest position, and install the arbor plate with its marked face toward the miter gauge.

Next, attach the dial indicator to the side of the notch in the block so that it projects from the end. Hold the block against the miter gauge, and position it so the tip of the indicator is on the centerline of the arbor plate. Now move the indicator toward the arbor plate so the hand of the indicator makes a couple of turns and comes to rest at the 12 o'clock position from your point of view. Clamp the bar to the miter gauge (see the photo at left below).

Using a pencil, make a mark on the arbor plate where the indicator tip contacts the centerline. Use a square to lengthen the mark to make it easily visible (see the photo at right below).

With one hand holding the miter-gauge assembly steady against the outside edge of the miter slot, use a fingertip to tap lightly a few times on the indicator stem and the face of the arbor plate to be sure they've settled into position, then zero the indicator (see the top photo on p. 76).

Marking the plywood arbor plate helps ensure a more accurate test. The tip of the dial indicator should be placed on the same reference point when checking the alignment from the front and back of the table.

Mounting a piece of Baltic birch plywood in place of the blade allows a precise check of whether the miter-gauge slot is parallel to the blade.

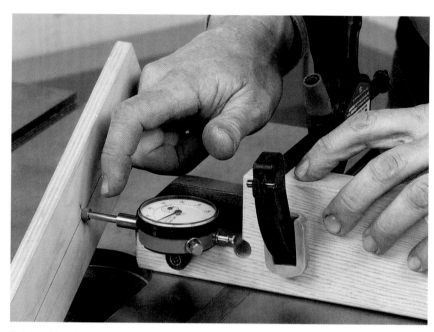

Before taking the reading, use your fingertips to tap on the arbor plate and dial indicator to make sure the indicator has settled into position, then zero the indicator.

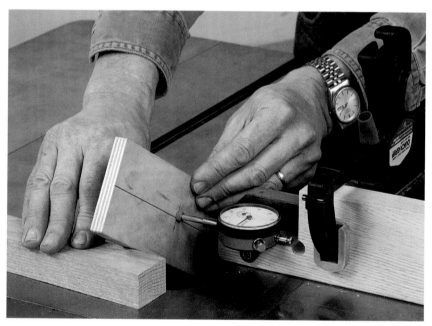

Take the reading at the back of the saw at the same center mark on the arbor plate. A small block may be needed to prop up the plate to bring the tip of the indicator on the mark.

Rotate the arbor plate to the opposite end of the opening, and move the miter gauge forward until the indicator tip again touches the center-line at the spot you marked earlier. You may need to prop the arbor plate

To align the miter-gauge slot with the blade on a cabinet saw, start by loosening the bolts that attach the top to the cabinet.

Once the table bolts are loosened, tap the top gently to shift its position, then recheck the slot-to-blade alignment with the dial indicator.

on a small block (see the bottom photo on p. 76). If the dial indicator doesn't still read zero, the miter slot isn't parallel to the plane of the saw-blade. To get an accurate reading, always apply light pressure to the miter gauge so that it stays against the outside edge of the slot.

Slightly flex the arbor plate with finger pressure to see which way brings you back toward zero. This will tell you which way the table needs to rotate to make the blade and the miter slot parallel.

For cabinet saws On cabinet saws, the top is attached to the base with bolts at each corner. When these are backed off, the top can be shifted to bring the blade and the miter slot into alignment. I find the easiest approach is to loosen three of the bolts and leave the fourth slightly snug to serve as a pivot (see the photo above). Tap the corner diagonally opposite the pivot bolt with a soft mallet or wood block (see the photo at right). If the table won't shift far enough, choose a different bolt for the pivot point.

Check your progress using the dial indicator and the arbor plate as you go. You'll have to reset the indicator to zero each time. When you are satisfied with the alignment, snug up the bolts, then recheck the setting just in case the top shifted as you tightened the bolts.

Contractor's saws On a contractor's saw, the trunnions that support the arbor are bolted to the underside of the table. To prevent the arbor assembly from shifting or dropping out of the trunnions, clamp the front and rear trunnions as shown in the photo at left on p. 78 before backing off the bolts.

Keeping It Together

When checking the alignment of the miter-gauge slot, you will need to move the dial indicator and the arbor plate from one end of the insert opening to the other and back again to take readings. The job will go faster if you move the indicator with one hand and the arbor plate with the other, keeping the tip of the indicator on the plate as you rotate it into position. By doing this, you won't have to draw back the tip of the indicator to reset it on the plate each time you switch positions.

To adjust the alignment on a contractor's saw, start by clamping front and rear trunnions. Bolts holding the trunnions in place are attached directly to the underside of the table.

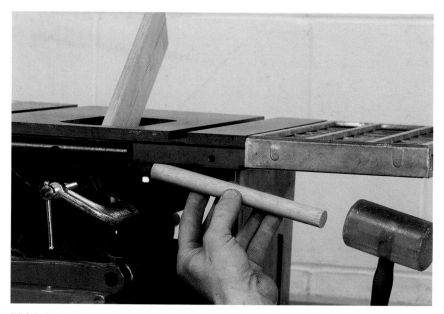

With bolts loosened, gentle taps using a hardwood dowel will shift the trunnions so the miter-gauge slot becomes parallel with the blade.

Bolts That Shift

One common annoyance when tuning up a machine is to get everything adjusted just right and then have parts shift out of line when you tighten the bolts. If that happens, check the underside of the bolt heads for burrs, paint, or rough spots and dress them using a file. Also check that the spot on the machine where the bolt attaches is smooth and paint free. The bolt should have a sturdy, burr-free, flat washer under the head in addition to any lock washer. Finally, grease the threads of the bolt and both faces of the washer before assembly. This is a good place for an antiseize grease.

With the trunnions safely clamped, loosen three of the four bolts holding the assembly to the bottom of the table. The fourth one, at the front of the saw, should be left slightly snug to serve as a pivot. Adjust the alignment by tapping the rear trunnion bracket to rotate the arbor assembly into better alignment using a mallet and hardwood dowel (see the photo above right).

Once the dial indicator shows the alignment is correct, snug up the bolts and reconnect the drive belt. If the weight of the motor throws off the alignment, you'll have to readjust the trunnions to compensate. Remove the clamps when you are done.

Benchtop saws The arbor assembly on a benchtop saw is suspended from a pair of small hinges bolted to the underside of the table. Loosen the bolts, which can be hard to reach, leaving one slightly snug to serve as a pivot point (see the photo on the facing page).

Adjust the miter-slot alignment by shifting the bracket that isn't serving as a pivot. Use the dial indicator to measure the alignment, then tighten down the bolts when it is correct. Recheck to make sure the position hasn't shifted.

Aligning the Rip Fence

A rip fence is properly adjusted when it is parallel to the blade. If it isn't, the ripped board will have a burned or wavy edge and the chances of kickback increase. The procedure for checking the alignment of the fence is similar to the method used to check the miter-gauge slot.

A pair of small hinges bolted to the underside of the table controls the miter-slot alignment on a benchtop saw. Reaching them can be a chore.

Because many original equipment fences have design flaws that make them unreliable and difficult to align, a dozen or more aftermarket rip fences are available (although not all of them are much of an improvement). In my book, the best design is the one by Biesemeyer®, of which there are now several clones.

If the outfeed end of the fence is toed in toward the blade, the stock is pinched against the side of the blade during a cut, which creates a host of problems (see the illustration at right). Initially, the board will be hard to feed. Friction will burn the cut edge, and the blade will overheat. When the leading end of the stock reaches the back of the blade, it will be struck by the rising saw teeth, causing a ragged edge and a high risk of kickback.

If the outfeed end of the fence is toed out, the symptoms are distinctly different (see the illustration on p. 80). The kerf in the board, sandwiching the blade, will lead the stock away from the fence. Attempting to force it back will create a wavy edge on the board. The offcut will drag against the side of the blade, leading to burning, chatter, and possible kickback.

Some rip fences are bolted to the head with two to four bolts, which when loosened allow the fence to be shifted into line. This style of fence is harder to adjust; you just have to keep trying to get it right. Some bolt-type fences lock onto the front rail only, while others lock onto both the front and rear rails. Adjustment procedures are somewhat different. In Biesemeyer-style fences such as that shown in the photo on p. 80, adjusting screws are located behind pads at the outer ends of the fence's head. Setting up this style of head is a joy. You just tweak the set screws until the fence is where you want it.

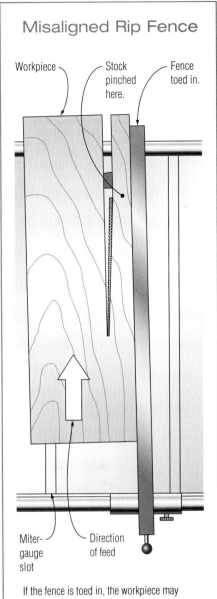

Misaligned Rip Fence

Workpiece

Stock pinched here.

Fence toed in.

Miter-gauge slot

Direction of feed

If the fence is toed in, the workpiece may become jammed between the back edge of the sawblade and the fence, burning the stock and increasing the risk of kickback.

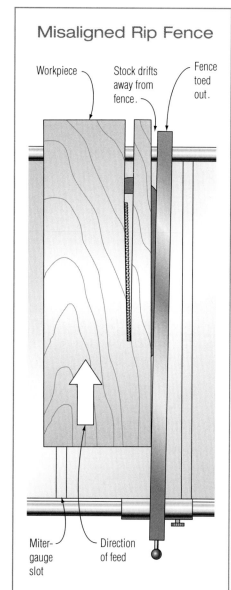

Misaligned Rip Fence

Workpiece

Stock drifts away from fence.

Fence toed out.

Miter-gauge slot

Direction of feed

If the fence is toed out, the front edge of the workpiece will drift away from the fence. Attempts to push it against the fence will cause binding and possible kickback.

In this Biesemeyer-style fence, set screws behind nylon pads on the head of the fence are used to adjust alignment.

CHECKING THE ALIGNMENT

Start the alignment procedure by unplugging the saw, removing the blade, and placing the fence on the side of the table you normally use for ripping. Raise the blade-height adjustment to just below its highest position, and install the arbor plate with its marked face toward the fence.

Attach the dial indicator to the end of the notch in the block as shown in the photo on the facing page. Holding the block against the fence, move the fence toward the arbor plate so the hand makes a couple of turns and comes to rest at the 12 o'clock position. Lock the fence in place, then slide the block along the fence until it crosses the centerline on the plate and mark the intersection. Using a square, lengthen the mark. Tap the indicator and plate a few times to make sure they've settled in and zero the indicator.

Next, pivot the plate to the other end of the opening and slide the block forward until the indicator tip again touches the centerline at the

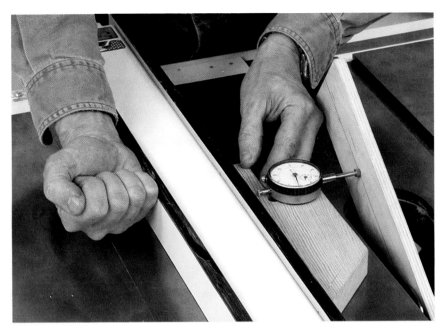

The rip fence should be perfectly parallel to the plane of the blade. Measure at the front and rear of the table with a dial indicator.

spot you marked earlier. You may have to raise the plate on a block to get the tip on the correct spot.

If the dial indicator doesn't read zero, the fence isn't parallel to the blade. When taking readings, lightly press the block against the fence and make sure that no sawdust is caught behind the block. Slightly flex the arbor plate with finger pressure to see which way brings the indicator back toward zero. This will tell you the direction the outfeed end of the fence will need to be adjusted.

ADJUSTING FRONT- AND REAR-LOCK FENCES
To adjust front- and rear-lock fences, unlock the head after you have taken your initial measurements with the dial indicator. Set the indicator aside, then one at a time, loosen the bolts holding the fence to the head and retighten them until they are lightly tensioned. Leave the fence free enough to be tapped into alignment with light mallet blows. Once you have the bolts adjusted, tap the outboard end of the fence to bring it closer to parallel to the blade.

Lock the fence to the bar. If the pull of the lock causes the head and bar alignment to shift, you'll have to tighten up the bolts before you do this. Move the dial indicator back in place and compare the front and rear readings to see if the fence is properly aligned.

Repeat the procedure until the indicator shows the fence is aligned, then fully tighten the bolts.

Rip fences on contractor's saws often need adjusting after loosening the bolts at the head.

The rip fence is parallel to the blade when dial-indicator readings at the front and back of the fence are the same.

A gentle tap is all that's needed to move the fence into alignment.

ALIGNING A SET-SCREW FENCE

If you have a set-screw fence, lift the fence off the front rail and flip it over on the saw table. Adjust either the left or right set screw to correct the fence's setting, then replace the fence and use the indicator to check the

Set screws adjust the alignment of the T-style fence.

alignment. Repeat the adjustment procedure until the fence is parallel with the plate. Everything should be this easy.

On this style fence, the set screws also affect how tightly the fence locks onto the rail. If adjusting for parallel has made the lock setting too tight or too loose, advance or retract both screws equally to get the lock setting right, then recheck the fence's alignment with the blade. It may need a slight adjustment.

Tuning Up the Blade-Tilt Settings

Properly adjusted, the sawblade will be perfectly square to the surface of the table when it is moved all the way to its vertical stop. A second stop sets the blade at 45 degrees for cutting miters. The quickest and most accurate test for setting these two stops is a dynamic one: You cut the end off a board and check it for square or 45 degrees.

To do this test, choose a 3-ft.-long, clear, straight-grained piece of wood about 1 in. thick and 4 in. wide. It can be either hardwood or softwood. If you choose pine, it should be a dense, tight-grained piece that will cut cleanly.

Carefully plane one face of the board flat, since this test will be reliable only if the board sits perfectly flat on the saw's table. Once you have a flat face, plane one edge straight and square to the good face (this edge will ride against the miter gauge when you do the test cuts).

Place the board on the saw table with the planed face down and the squared edge against the miter gauge, then mark the exposed face of the board with a prominent V pointing toward the face of the gauge. Make all of your test cuts with the board in this position.

The only tools you will need are a square that you know is accurate and a few small regular wrenches or Allen wrenches for adjusting the stop bolts.

Mount a good-quality crosscut or combination blade on the saw for this procedure. The smoother the cut end of the board, the easier it is to judge the board's fit against the square.

SETTING THE 90-DEGREE STOP

The stops at 90 degrees and 45 degrees allow you to return to these settings quickly with a fair degree of accuracy. To check the 90-degree setting, start by cranking the handwheel until the tilt mechanism is against the vertical stop. Lock it and place the miter gauge in the slot you regularly use for crosscutting (see the photo at left below).

Take a test cut off the end of the test board, trimming off about ½ in. of stock. Make sure the V on the board is visible and pointing toward the miter-gauge face. To check the cut for square, place the beam of the square on the planed (unmarked) face of the board and hold it up to the light to see if there is a gap between the freshly cut end and the square's blade.

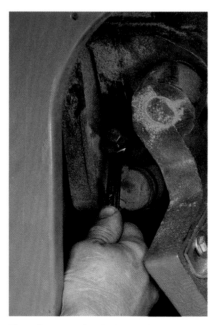

Prepare to check the accuracy of the stop by setting the saw to the vertical position and placing the miter gauge in the slot you regularly use for crosscutting.

The adjustment for the blade-tilt stop is generally tucked up inside the saw in a hard-to-reach spot. It can sometimes be reached through the slot in the front of the cabinet.

Once you have the 90-degree adjustment set, adjust the tilt-stop pointer.

If the cut isn't square, adjust the stop bolt and tighten its locknut, then take another test cut. On most saws, the stop bolt is under the table and awkward to reach. Sometimes it can be reached through the slot in the front of the cabinet (see the center photo on the facing page). On a few saws, the stop bolts are set into the tabletop and are easily adjusted with Allen wrenches. These bolts don't use locknuts; they are locked by a nylon insert that prevents them from shifting.

Once the saw is cutting a square end on the test board, set the pointer against the angle scale to zero. For accurate readings, the pointer's tip should be close to the scale. You may have to bend it slightly to get a better fit.

SETTING THE 45-DEGREE STOP

You can use essentially the same procedure for checking the 45-degree stop. An accurate check, however, requires two test pieces cut at 45 degrees, not one. First, tilt the blade until it reaches the stop, lock it, and place the miter gauge on the side of the table the blade tilts toward. Take a test cut, trimming about ½ in. off the board, then make a small V mark on the board, pointing toward the end you just trimmed off (see the photo below).

Slide the test board over 4 in. and take another cut. Setting the 4-in. piece aside for a moment, again make a mark on the test board pointing toward the end you've just cut. Then take the 4-in. cutoff and the main board and arrange them to make a corner with the V's pointing at each

Mark the end of the 45-degree test cut with a small V.

The two 45-degree test pieces should mate at exactly 90 degrees.

Tilting the blade back toward vertical may be necessary to access the 45-degree stop screw.

other at the inside corner of the joint. Place a square on the outside of the corner and check if it is a precise 90-degree corner (see the photo at left above).

If the test corner isn't square, adjust the stop screw, lock it, and make a new test joint. On many saws, the 45-degree stop screw can be difficult to reach. If so, try tilting the blade back toward vertical to improve your access to the screw and then return it to the stop for a new test.

Checking the Miter Gauge

Of all the steps involved in tuning up a table saw, adjusting the stops on the miter gauge is the easiest. The gauge should help you make precise 90-degree crosscuts and 45-degree miters. The simplest and most accurate method of aligning the gauge is to make test cuts and then check the cut end of the board with an accurate square.

Before adjusting the gauge, make sure that any shopmade wood facing is straight, square to the table, and solidly attached to the gauge's head casting. Often the bar will seem loose in the miter slot but this rarely has a significant effect on accuracy because the bar will settle into the same orientation in the slot each time a cut is made.

Almost all original equipment fences are adjusted by turning stop screws that bear against a small hinged tab in the trailing edge of the bar. This style of gauge is frequently thrown out of alignment because the tab bends easily. Handle the gauge with care and check it before making critical cuts.

Loosen the set screw to adjust the pointer on the miter gauge.

If adjusting the stops on the gauge won't produce smooth, square crosscuts, it is likely that the miter-gauge slot is not parallel to the blade. The only solution is to reposition the table as outlined in "Aligning the Blade and Miter-Gauge Slot" on p. 73.

If you are only going to check the miter gauge for square cuts, you'll need one board around ¼ in. thick, 3 in. to 4 in. wide, and 2 ft. long. If you also want to set the 45-degree stops, you will need a second board the same size. The stock can be either hardwood or a tight-grained softwood with no large knots. You should plane the stock flat and even in thickness and carefully joint one long edge straight and square. Mark V's on both wide faces of the board pointing to the jointed edge. Always make your test cuts with this edge against the miter gauge.

Place the gauge in the slot you regularly use for crosscutting, then place the test board against the gauge with the good edge against the face and trim ½ in. off the end. Using the square, check the edge and make the appropriate adjustment on the stop bolt. When the cut is square, adjust the pointer to 0 degrees (see the photo above).

To adjust the 45-degree setting, turn the head of the gauge toward the blade and lock it against the stop. Take one of the boards and trim its end off, making sure you keep the good edge against the face. Make the same cut on the second board. When the two boards are butted together, they should form a 90-degree corner. Adjust the stop screw, if needed, until it is exactly 90 degrees.

Check Miter Slots for Parallel

The two miter slots in the saw's table may not be exactly parallel or precisely the same width. If you tune up the gauge in the left-hand slot, it may need to be readjusted when you use it in the right-hand slot.

The Thickness Planer

Accurate joinery is only possible when both faces of a board are flat and parallel to each other. A jointer can flatten the face of a board, but only a thickness planer can make both faces parallel and the stock even in thickness. A thickness planer, along with a jointer, is essential for transforming roughsawn planks into stock ready to be made into furniture.

Until fairly recently, thickness planers were the heavyweights of the woodworking shop. A machine capable of planing a board 20 in. wide might easily weigh half of a ton; even small machines weighed several hundred pounds. In the 1970s, lighter and less expensive planers became available and rapidly gained popularity in both large and small cabinet shops. These benchtop machines are now made by a number of manufacturers.

Anatomy

The operating principle of a thickness planer is simple. The planer's cutter-head is locked at a fixed height over a table. With one side already planed flat, a board passed under the cutter is reduced to a uniform thickness.

The cutting head on the smallest machines holds just two knives, whereas most large machines have three or four. Some large industrial machines have a series of short, staggered knives that function like a spiral cutter. Whatever type of blade the machine has, it will work properly only when the blades are sharp and precisely aligned.

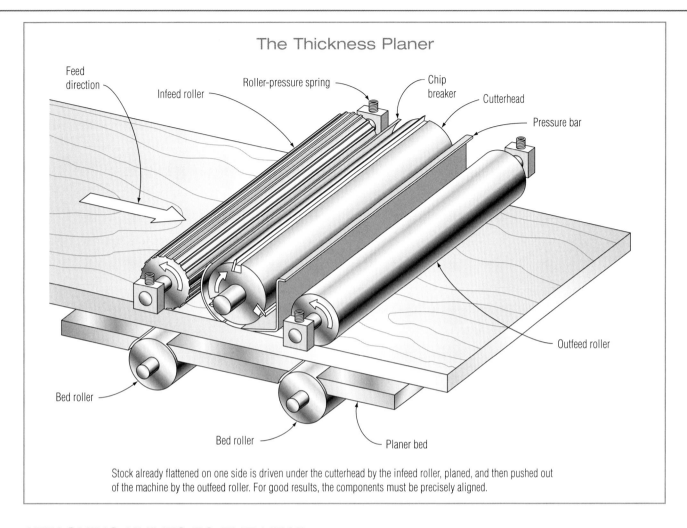

The Thickness Planer

Feed direction

Infeed roller

Roller-pressure spring

Chip breaker

Cutterhead

Pressure bar

Outfeed roller

Bed roller

Bed roller

Planer bed

Stock already flattened on one side is driven under the cutterhead by the infeed roller, planed, and then pushed out of the machine by the outfeed roller. For good results, the components must be precisely aligned.

ATTACHING KNIVES TO THE HEAD

There are three general approaches manufacturers have taken to holding and adjusting knives in the cutterhead: a preset alignment typical of disposable blades, spring-backed knives set with a simple jig, and jackscrew-adjusted knives that are often set with a dial indicator supplied by the manufacturer. All three methods work well, although the preset disposables are, of course, the quickest to change and are virtually foolproof in their alignment.

There also are several ways of locking the knives into the head. On older and larger machines, a gib bar securing the blade is locked into a groove in the head by bolts spaced a couple of inches apart. These tend to be fussy to adjust. The lock bolts can be awkward to reach, and the blade will often creep out of position as the bolts are tightened (see the top photo on p. 91). Portable machines often use a flat mount that clamps the blade under a broad plate. This type of blade attachment is easy; the bolts are large and easy to reach and the blade won't shift as the bolts are tight-

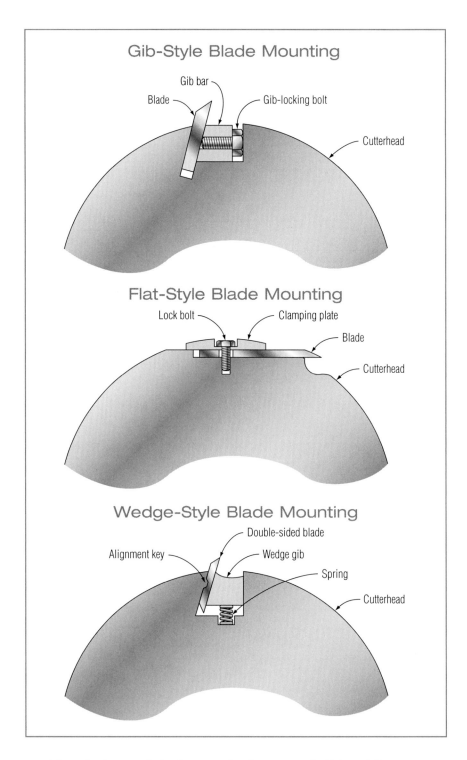

Gib-Style Blade Mounting

Gib bar

Blade

Gib-locking bolt

Cutterhead

Flat-Style Blade Mounting

Lock bolt

Clamping plate

Blade

Cutterhead

Wedge-Style Blade Mounting

Double-sided blade

Alignment key

Wedge gib

Spring

Cutterhead

ened (see the bottom left photo on the facing page). Gib and flat-mount arrangements are used for both disposable and standard blades.

A third method, which is good only for disposable knives, uses wedges set by centrifugal force to hold knives in place (a well-known brand of this

Common gib screws are awkward to tighten because the small bolt heads are difficult to reach.

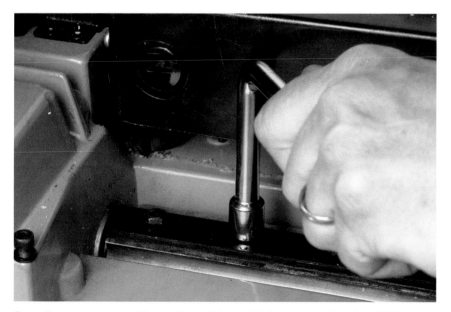

Generally seen only on smaller machines, flat-mount knives are easy to set and tighten.

Wedge-lock knives are easy to change. Wedges are tapped back against small springs to release the knives, then with new blades in place, the wedges pop back as soon as the machine is turned on.

type is TERSA). Wedged knifes are the simplest and fastest to change, but this system is typically available only on larger commercial machines (see the photo at right). The only small planer I know of that uses this approach is the Inca jointer/planer, a combination machine designed for a small shop.

Disposable knives are standard on many portable planers and are available as an option on heavier industrial machines. Most disposable knives are a thin strip of high-speed steel sharpened on both edges. A lengthwise groove or holes for alignment pins precisely position the blade in the planer's cutterhead. Disposable blades are often more economical than standard blades when setup time, resharpening cost, and quality of the cut are factored in.

KEEPING STOCK MOVING AND CHIP FREE

All planers use feed rollers on both the infeed and outfeed sides of the machine. Large planers have corrugated steel rollers for extra traction on the infeed side and nonmarring steel feed rollers on the outfeed side. Portable machines use rubber-covered rollers on both sides of the cutterhead. To reduce friction between the table and the stock, most large machines have a pair of unpowered bed rollers set in the table directly below the feed rollers.

To prevent tearout, large planers have a chip breaker that rides on the stock on the infeed side of the cutterhead. By bearing down on the wood a fraction of an inch away from the knives, the breaker forces the chips raised by the knives to break off instead of running along the grain and tearing chunks from the surface. Portable planers don't have chip breakers. Instead, they rely on higher cutterhead speed, limited depth of cut, and a slow feed rate to limit tearout.

Large machines also have a pressure bar on the outfeed side that holds the board tightly against the table. By preventing the end of the board from rising, the pressure bar minimizes snipe, the small but annoying divot

Resharpening Disposables

Some disposable knives are designed with enough extra metal in the blade to allow them to be resharpened two or three times before being retired. If you attempt this, remember that the grinding must be very precise to keep the blades matched to each other in width.

The head of this portable planer rides on four posts, adjusted by the crank on top of the machine.

On large planers, the table moves up and down to adjust for stock thickness.

Portable planers achieve a low-speed drive through a gear box attached directly to the motor. Only the final chain drive will need service.

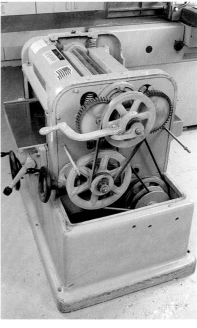

Standard V-belts connect the motor to the cutterhead on this large machine. Belts also step down the speed for the drive rollers, but the final low-speed drive is chain driven.

that can appear on the first and last few inches of a planed board. Pressure bars aren't used on portable planers. An outfeed roller mounted close to the cutterhead is designed to control snipe.

Most smaller planers adjust for stock thickness by allowing the cutter-head assembly to move up and down on two or four posts (see the photo at left on the facing page). In large, cast-iron planers, the head is normally fixed and the table moves up and down on slides cast into the machine's frame (see the photo at right on the facing page). In both designs, the position of the table or head assembly is adjusted by jackscrews.

DRIVE SYSTEMS VARY BY MACHINE SIZE

Large planers are powered by conventional electric motors (sometimes using three-phase current) that drive the cutterhead with V-belts (see the photo above right). To get a slow-speed drive for the feed rollers, a series of step pulleys or a transmission is used. The final stage is a chain drive to the rollers. Most large machines have a simple lever-operated clutch in the roller drive system and an adjustable feed rate.

Portable planers use a compact, high-speed universal motor for power. One end of the motor's shaft drives the cutterhead with a cog-belt drive. The opposite end of the motor connects directly into a gear box that is part of the motor housing. The gears provide a low-rpm drive to the rollers. The final drive is a chain and sprocket assembly (see the photo above left).

Troubleshooting

Thickness planers typically misbehave in one of three ways: snipe at one or both ends of the board, difficulty in feeding the stock, or a poor finish on the planed surface.

SNIPE

A snipe is a slightly deeper cut in the leading and trailing few inches of a board (see the photo below). In most situations, the cause is simple: The stock lifts off the planer's table, allowing the knives to cut too deeply. When both feed rollers are pressing down on the board, it can't lift, so sniping occurs near the ends of the board when only one roller is applying pressure. Large planers have chip breakers and pressure bars close to the cutterhead to further limit lifting of the stock. Adjusting those parts is critical to preventing snipe.

Snipe can result when the long end of the stock is not supported. The weight of a long board can easily lever its leading or trailing end off the

Check the Stock

Not all sniping is due to faults in the machine. Poorly prepared stock also can cause the problem. When the back of the board—the side that goes against the table—isn't flat, or when a board has a lengthwise bow in it, sniping may result.

Sniping leaves a small step on one or both ends of a board. It can be reduced or eliminated by changes in procedure and careful tuning of the planer.

Locks holding the planer head in position during a cut can reduce sniping.

table, causing a deeper cut. As the first and last portions of the stock are being fed, the opposite end of the board should be held level with or slightly higher than the planer's table. Using infeed and outfeed tables or roller stands is a good solution. The problem occurs more often on portable planers.

If your machine has bed rollers, they should be set at least a few thousandths of an inch above the table's surface. If the bed rollers are low, the pressure of the feed roller can flex the board downward into the bed-roller slots, causing the board's ends to curl upward slightly and creating sniped ends. If the bed rollers are set too high, you can get a reverse snipe, where the first or last few inches of the board will have a slightly higher step.

Some portable planers, however, have a design flaw that can cause sniping. As the leading edge of the stock begins to pass under the outfeed roller, the power head can flex slightly upward due to the added pressure. This lifts the cutting head and causes a step in the board's surface. When the trailing end of the board clears the infeed roller, the opposite motion occurs. The head drops when the pressure is off the roller, sniping the trailing end of the board. To prevent this, many second-generation machines were designed to be stiffer and had head-locking mechanisms added to eliminate movement of the head once the height was set (see the photo above).

If your portable machine doesn't have a head lock, a certain amount of snipe may be inevitable, especially if wear has loosened its fit on the guideposts. If the machine has a locking mechanism, make sure it is working properly. In large machines, where the table moves instead of the power head, a similar problem can occur if the table's gibs are loose, allowing the table to rock as the stock is fed.

STOCK WON'T FEED SMOOTHLY

There can be many reasons why a board won't feed smoothly through the planer. But before looking for a complex mechanical problem, it always pays to check the condition of the rollers and table. Dirt and pitch on these parts is usually the cause.

The table surface should be as friction free as possible. A good first step is to wipe down the table with solvent on a rag. Use a scrub pad if necessary to remove heavy buildups on cast-iron tables, but avoid the use of abrasive pads on the chrome-plated tables of portable planers because the finish can be damaged. If the table is rusted, polish it with emery paper and oil and follow up with wax or a coating of TopCote®, a lubricant made for the job. If you use a spray, be careful to keep it off the feed rollers so they don't get slippery.

If the planer has bed rollers, make sure they're clean and revolve freely. A bent roller will cause the stock to advance erratically. If a roller is higher on one end than the other, the stock will skew sideways as it is fed.

While you are cleaning the table, it pays to clean the feed rollers too, even though they're awkward to reach. Use a solvent and rags, going over the rollers a couple of times to make sure they are free of dust or oil that might make them slip. If the rollers are rubber covered, take a close look at them. You should be able to press the tip of your thumbnail into the rubber easily. If the rubber is glazed and no longer resilient, the roller will have trouble gripping the stock and should be replaced.

While working under the cutterhead, check the bearing surfaces of the chip breaker and pressure bar, if your machine has them. They should be clean and free of burrs. You can also wax them, but be careful to keep wax off the nearby rollers. Rollers that are set too high or too low will cause feeding problems, as can misaligned chip breakers and pressure bars. To set these components, see "Tuning Up the Cutterhead" on p. 103.

Dull knives also can lead to feeding problems. If knives are dragging rather than cutting cleanly, the rollers will slip trying to force the stock under the cutterhead. Try feeding narrow stock at the far left or right edge of the table where the knives are probably sharper. If the slippage disappears, the center sections of the knives are likely dull.

Occasionally the drive line will cause trouble if a V-belt is slipping or a pulley or chain-drive sprocket is loose. If the stock isn't feeding, glance under the head (from the side of the machine for safety) while the power is on. If the feed rollers aren't turning, the problem is in the drive line.

ROUGH FINISH SURFACE

In addition to causing feeding problems, dull knives create burned and glazed areas on the surface of the stock that will lead to problems when gluing and finishing. The next time you install new knives, plane and set aside a few sample boards in both hardwoods and softwoods. Later, when the knives have become duller, you'll have a reference to judge the degra-

Pitch buildup on the planer's table is the most common cause of feed problems. To prevent the problem, clean and wax the table occasionally.

To support the dial indicator during a tune-up of your thickness planer, you should construct a small base for it. The one shown below is made from MDF, but you could also use plywood or hardwood. The indicator is attached to the base with a ¼-in.-dia. bolt and a wing nut. If your planer does not have bed rollers, the ½-in. hole in the bottom piece of the base won't be needed.

The upright part of the base is attached to the bottom with two drywall screws, which should be countersunk. The base stands on three legs, which are ordinary flathead sheet-metal screws, ¾ in. long, that protrude about ¼ in. (see the bottom photo). The legs prevent dust or wood chips from being caught beneath the unit, which would cause a measurement error.

If your planer has bed rollers, the legs have another advantage. They allow you to adjust the dial indicator up or down so it can be zeroed out on the plane bed with the zero at the 12 o'clock position. Although this isn't absolutely necessary, it makes the dial easier to read and may help to prevent errors. Once the indicator is set this way, you can measure and adjust the height of the rollers more easily.

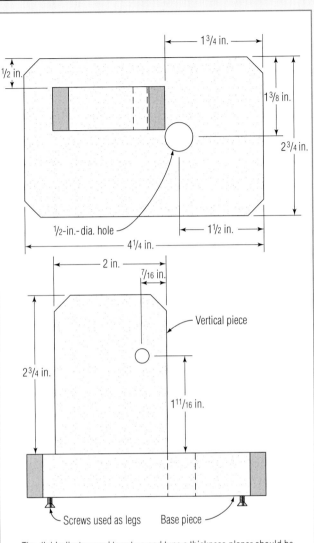

The dial indicator used to set up and tune a thickness planer should be mounted on a base made from a flat, smooth material, such as MDF.

Mounting a dial indicator on this shopmade base makes it easy to measure and adjust the height of the planer's feed rollers, pressure bar, and cutterhead.

dation of surface quality. Under a magnifying glass, the difference can be dramatic.

A knick in the knives will leave a raised line down the length of the board. If the knives are otherwise usable, loosening and resetting one knife a fraction of an inch sideways will eliminate the ridge. If one knife is set higher than the others, the stock will have a pronounced ripple in its surface. This effect is most noticeable at high feed rates. A cutterhead with a high knife will also sound different; the high knife will make a slightly louder snicking sound that is distinctive once you've learned to recognize it.

A planer can make a lot of chips in a hurry, and if they aren't cleared away they can wind up getting caught between the spinning knives and the surface of the board. Chips caught under the blades cause numerous small dings in the surface, which is more pronounced in softwoods or when taking a heavy cut (see the photo at left). On portable planers, this problem often occurs when an accessory dust hood has been attached but is not hooked up to a dust collector. The added restriction of the hood blocks the smooth flow of chips. If the problem occurs on a planer that does have dust collection, check the system for a clog or restriction that is cutting the airflow.

Setup and Maintenance

Large cast-iron planers meant to stand directly on the floor don't require much in the way of setup beyond being leveled and supported evenly on all four corners. Generally, they do not need to be bolted down. Because planers generate a huge amount of chips, they should be located close to a central dust collector with as few restricting bends in the ductwork as possible.

You should bolt portable planers on a stand for stability and to bring them up to a practical height. The table should be about 24 in. off the floor (see the photo on the facing page). When possible, hook up a portable planer to a 4-in.-dia. dust-collection hose. If you are using a 2½-in. shop vacuum for chip removal, keep the hose short and eliminate any tight bends. If you aren't going to use any dust pickup on a portable machine, leave off the accessory dust-collection hood.

Large planers use a V-belt drive off the motor that should be checked occasionally for looseness and wear (see the photo at left on p. 100). Portable planers use a short, cogged belt to drive the cutterhead. This type of belt typically has a very long life and should not need tightening.

If you have a large machine with a gear box, you should drain and replace its oil occasionally, once a year for frequently used machines. The small gear boxes on portable planers are sealed and don't require oil changes. If you don't have a service manual that specifies oil types, 80W-90 gear oil meant for car transmissions should work well. Without a manual, knowing how much oil to put in the gear housing can also be a problem.

When chips are not removed quickly enough, they can get caught under the knives, leading to these characteristic dings in the planed face of the board.

For safety and convenience, portable planers should be mounted on a stand roughly 24 in. high.

If the oil is filled through a plug in its side, the gear box is usually filled until the oil reaches the bottom edge of the filler hole. If the box is filled through a top-mounted plug, you should not fill it all the way to the top; look for a cast-in oil-level mark on the outside of the box. If there is no mark, leave the oil an inch or so below the inside edge of the filler hole.

On most planers, the feed rollers are powered by a chain drive, which you should clean and reoil occasionally. A spray penetrating oil and some rags do a good job of removing the old grease and accumulated sawdust. Use a heavy oil to lubricate the chain, or use a spray lube designed for chain drives.

Almost all of the following information on tuning up a planer applies to full-sized cast-iron machines, which have many components to adjust. If you own a portable planer, you'll have considerably less to do. The sections on adjusting the chip breaker and pressure bar won't apply, for example, because these small machines don't have them. Nor will the sections on adjusting roller heights and pressures; although portable machines have these parts, they can't be adjusted.

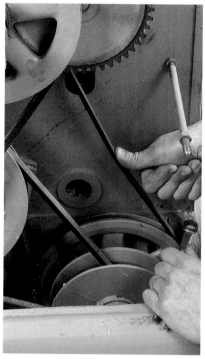

When properly tightened, a belt should flex only a fraction of an inch under firm finger pressure.

Be sure to clean and lube the table slides on planers before adjusting the gibs.

Table Adjustments

Thickness planers whose heads or tables move up or down on posts—and this includes almost all portable planers—generally don't have a way of adjusting how tightly the head or table assembly fits on the posts. These machines often make up for any play in the fit with a locking mechanism. A small number of machines, however, have gibs or split bushings that can be adjusted. In this case, tighten them to remove any play but leave them just loose enough to slide smoothly. If the machine has a mechanism for locking the head or table on the posts, make sure it locks solidly. The locking mechanism should have some adjustment for tightening the fit.

In large planers, the table typically runs up and down on rectangular cast-iron ways machined into the side castings. You can adjust the gib screws to eliminate any play between the moving component and the supports while allowing the table to slide smoothly (see the photo at right on the facing page).

CHECKING FLATNESS

Two surfaces in a thickness planer must be flat for the machine to work properly. The first is the table that the stock rides on. If the table is bowed or twisted, the stock won't feed smoothly and the finished board will be uneven in thickness. The second critical surface is the edge of the pressure bar. If the bar doesn't bear down evenly, the leading and trailing ends of the board can lift off the table and cause snipe. Part of the tune-up will deal with adjusting the pressure bar, but it can't be properly adjusted if it isn't straight and smooth to start with.

If the planer's table isn't flat, boards will be uneven in thickness. Ideally, the table will vary no more than 0.005 in. anywhere along the straightedge.

Check the table for flatness using either a reliable straightedge or a test bar made up according to the instructions on pp. 26–31. Check across the length, the width, and both diagonals of the table. If your machine has bed rollers that interfere with the straightedge, you may be able to remove them. In some machines, the rollers can be set so they are below the surface of the table. If not, use three identical pieces of shim stock to lift the straightedge over the rollers. If the table is out of flat, use feeler gauges to measure the error.

For most work, an out-of-flat measurement of 0.005 in. over 12 in. is tolerable. Beyond that, you're probably going to have joinery problems later as you use the wood. If a cast-iron table is out of flat, you will either need to replace it or remachine it to correct the problem. On portable planers, you can fix an out-of-flat table by layering aluminum-foil shims between the base casting and the sheet-metal working surface of the table.

To check the pressure bar, you will need a reliable straightedge, which doesn't need to be as long as the bar. By sliding a 12-in. straightedge back and forth, you can reliably check a 16-in. or 18-in. bar. Begin by running your fingers down the length of the bar, being careful of the knives in the nearby cutterhead. (The power should be locked out whenever you're working on the machine.)

As you run your fingers along the bar, feel for burrs and low spots, then use the straightedge to check the bar for overall straightness. A light shining from behind the bar will show any gaps between the bar and the straightedge. If the bar isn't smooth and straight, with the gaps less than 0.002 in., remove it from the machine and use a file to straighten its bearing edge and round its leading edge to a smooth profile.

EXTENSION TABLES

On portable machines, the extension wings should line up with the table under the cutterhead. The inner edges of the wings, where they are attached to the machine, should be even with or slightly below the height of the main table. If the wings are high, they will cause sniping and jamming problems. But there typically is no height adjustment here; if the alignment is off, you will have to change or modify the wing's pivot bolts.

The height of the outer end of each wing is adjusted by bolts mounted in the planer's frame near the wing's pivots. To adjust the wings, raise the power head to its highest position and place a builder's level through the machine and across both tables. Apply moderate down pressure on the level and adjust the bolts on each wing to give the wing a slight up-tilt (see the photo on the facing page).

When the wings are properly adjusted, the level will be parallel to the main table and raised about $\frac{1}{32}$ in. off the surface. Move the level from one side of the machine to the other, adjusting the two bolts aligning each wing until the level clears the table evenly all the way across. A slight upward tilt compensates for any downward flexing of the wing caused by

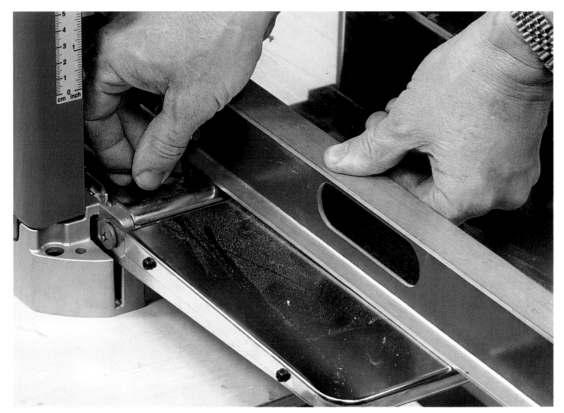

Fold-down tables on a portable planer should be adjusted to have a small up-tilt on their outer edges to compensate for their flexibility.

the weight of the board. In addition, it forces the leading and trailing edges of the board tightly against the table, reducing snipe.

Tuning Up the Cutterhead

Knives must be installed in a planer's cutterhead precisely. If one knife is higher than the others, it will dull rapidly and a planed surface will be rippled. To work their best, knives should be set evenly, with height varying by no more than 0.001 in. across their width and from blade to blade. You should clean and lubricate the knife-adjustment and locking components before installing new blades.

Begin tuning up the head by removing one knife and all its associated parts. To prevent distorting slotted-style cutterheads, you should remove and replace knives one blade at a time, locking in each new blade before moving on to the next one.

If you do need to remove all of the knives at once, gradually loosen the lock bolts of the knives in progressive steps, and reinstall the knives in the same way, gradually tightening all of the bolts as you work your way around the head several times. When removing all of the knives in the

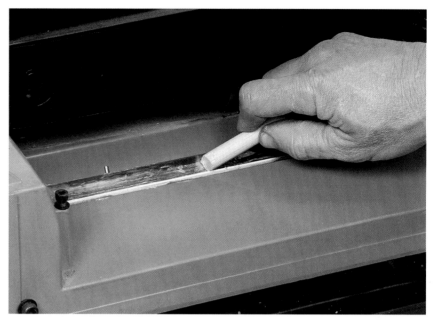

Before replacing a blade, always clean the slot or flat the blade mounts in. A dowel cleans the rounded profile of this chip-breaker groove effectively.

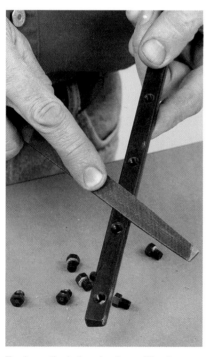

To clamp the knives in place without shifting, the clamp bars must be smooth and straight and the lock bolts must be burr free.

head at once, also remove the lift springs or jackscrews. Otherwise, they will drop out when you rotate the head to bring another knife to the top position. To keep track of parts and which blades have already been changed, use a marker to number each knife location on the head before unbolting anything.

Once you've removed the knife, use a solvent with scrub pads and scrapers to remove any accumulated pitch on the head. If the head is badly gummed up with pine pitch, use turpentine, which is an effective solvent. You can clean out round-bottomed chip-breaker grooves by using the end of an appropriately sized dowel as a scraper (see the photo above). A powered or hand-held wire brush is fine for cleaning up steel parts, but don't wire-brush aluminum because aluminum is too soft.

To remove burrs, lightly pass a file over the machined faces of the flats or grooves in the head. The surfaces of the clamping plates or the lock bars that bear against the knife blade should be smooth and flat. True them up using a file or by lapping them on sandpaper on a flat surface (see the photo at left).

The wrench flats on the lock-bolt heads should be flat and square, so clean up any rounded and burred heads using a file. Check that a wrench will slip onto all of the bolt heads in any orientation, and file down the bolt heads to fit snugly on just one size wrench. If the original wrench doesn't fit well, try a different one.

On slotted heads, the lock-bar bolt heads bear against the side of the knife groove to wedge the knife in place. Bolt heads should have a slight crown or conical point centered on the axis so they can be tightened without shifting the knife out of position. If a bolt head is burred, flat, or off-center, reshape it with a file or stone. If the bolt head needs extensive reshaping, keep it symmetrical by chucking it in a spinning drill while filing.

Polish the screw heads using fine emery paper so they will turn as friction free as possible against the side of the groove. Be sure to thoroughly wash off all the grit left behind by the polishing. If the debris gets onto other parts, it will cause binding and rapid wear.

Next, check that the bolts thread easily into the bars or the head itself. Burred threads on bolts can be cleaned up using a small triangular file; internal threads will need a tap to clear them. To aid in tightening, the threads of all bolts should always be lubricated with a small amount of grease during final assembly. Use a rag dampened with penetrating oil to wipe down the knives and the lock bars or clamping plates as you install them, leaving a very slight film of oil. The oil will reduce pitch buildup and enable the parts to move smoothly during adjustment.

Springs or jackscrew components that fit down into the head behind the blades should slide smoothly in their sockets and the screws should turn easily. Springs especially can get misshapen and bind. Always check, clean up, and lightly lubricate these parts before you install new knives.

Occasionally, parts in the head may need to be replaced. For safety and to preserve the balance of the head, always try to get replacement parts from the machine's maker. If you can't get original equipment parts, use high-strength bolts, normally called grade 8, and replace them and the springs in opposing pairs or full sets to keep the head in balance.

Sharpening and Installing Blades

In addition to being sharp, planer knives must be perfectly straight and even in width. Many disposable blades can be lightly resharpened one or more times, but they must be carefully ground to keep the set perfectly matched in width because their height can't be adjusted.

Since a precision grinder is an expensive tool that requires a fair amount of practice and skill to run, small and mid-sized woodworking shops often find it more economical to have a professional blade-sharpening service do the work or to use disposable blades. Small, low-speed wet grinders, typically with horizontal wheels, are sometimes advertised as being capable of grinding planer blades, but their performance is marginal at best. On these low-end machines, the carriage guiding the blade wobbles and flexes far too much to maintain the accuracy required for long planer blades.

Before installing a set of knives sharpened by an outside service, inspect them carefully. First, sight down each blade's length to make sure the blades aren't bent. Next, place the sharpened edge against your best straightedge and hold it up to the light. There should be no gaps, espe-

Maximum Blade Life

Check that your sharpening service is grinding away only the minimum amount of metal needed to restore the edge on your blades. If your service is removing more steel than necessary, you're going to be forced to buy replacement blades more often. I've noticed that some grinding shops are definitely more heavy-handed than others.

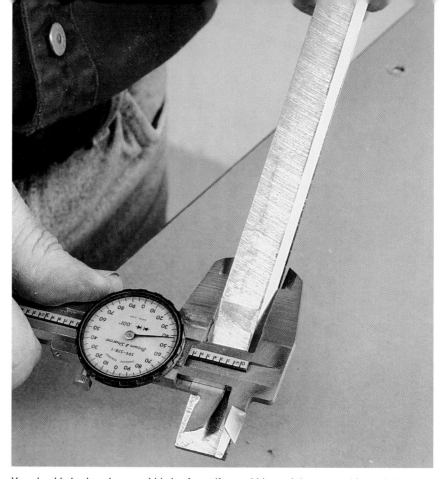

You should check resharpened blades for uniform width, straightness, and burrs before installing them.

cially at the ends where a poorly sharpened blade will often be rounded inward. Visually, the ground surface of the bevel should be consistent from one end of the blade to the other. Once again, most problems show up near the ends.

Using calipers, measure the width at the middle and at both ends of each blade. The width should be the same down the full length of the blade. To maintain the balance of the cutterhead, all of the blades in the set should be within a few of hundredths of an inch of each other in width. To line up properly when installed, reground disposable blades should be within a thousandth of an inch of each other in width.

The blades should be burr free. If they aren't, clean them up with a very fine stone, being careful not to round over the edge. The blades should not be discolored—that indicates that they were overheated when they were sharpened. A blade that has lost its temper won't hold an edge.

CHANGING PLANER KNIVES

Because there are many different jigs and methods of adjustment and locking for planer knives, it isn't possible to offer instructions that apply universally to all machines. The information in this section will guide you through the process in some detail, but you also should refer to the instruction manual that came with your machine for step-by-step procedures and the proper amount of knife exposure.

If you haven't done so already, read through the previous section on setting up the cutterhead. Knife changes will go faster and be more accurate if you take the time to make sure the cutterhead's lock and adjustment components are working properly.

TIGHTENING LOCK BOLTS

I'm covering this topic first because it is the one step in installing blades that is common to all machines. The best procedure for tightening blade-lock bolts, once the blade height is set, is to start by snugging up the bolts closest to the ends of each blade. Next, gently tighten the center bolts and, finally, the intermediate ones, alternating between the bolts on the left and right sides of the blade.

Once the bolts are all snug, go back and progressively tighten them using the same pattern in two or three steps. Using this procedure will usually eliminate the tendency of blades to shift out of line as they're locked in place. If a knife does shift, the problem is most often due to rough, poorly lubricated bearing surfaces on the bolt heads or warped lock bars or clamp plates.

INSTALLING ADJUSTABLE KNIVES

For adjustable knives, install a new blade, lightly snug up all of its lock bolts, and back the bolts off a quarter to half a turn. With the bolts backed off, the knife should move freely. Retighten the two bolts nearest the ends of the knife, then very slightly loosen these two bolts so the knife is again able to move but with just the barest amount of free play. By minimizing the free play, height adjustments will be more accurate and less prone to shift when locked down.

When adjusting knife height, work only on correctly setting the height of the ends of the blade; as long as the blade is straight, the middle will automatically be at the proper height.

INSTALLING SPRING-BACKED BLADES

Spring-backed blades are set using a jig supplied with the machine. The better jigs are double-ended, contacting both ends of the blade at once close to the outermost locking bolts. A variation on the spring-backed blade is a head without springs that depends on a magnet in the jig to lift the blade and hold it at the proper height. Magnetic jigs are a bit fussier to work with because the blades can easily be jarred loose before they're tightened.

For accuracy, always check that the contact points of the jig are free of sawdust or pitch that would throw off the height setting. Magnetic jigs are often thrown off by metal chips clinging to the magnet's face. To eliminate these chips, position the jig on the cutterhead and then lift it back off and

Don't Overtighten

Many people overtighten the lock bolts because they worry that a knife might fly off. But repeatedly overstressing the lock bolts will reduce their strength, which can lead to failure, creating the very situation they seek to avoid. The bolts should be firmly tightened, of course, but not so much that the wrench is damaging the bolt heads or the bolts are nearly impossible to back off at the next blade change.

Before tightening spring-backed blades, check that they're moving freely. Press the blade with a length of dowel and make sure it springs back against the jig.

check the faces of the magnets. This procedure will catch any steel chips that were down in the machine or clinging to the new blades.

To install knives with a double-ended jig, place a knife in its slot and press it back into position with the jig, being sure to locate the jig over the knife as the instruction manual indicates. To check that the knife is seating properly, press it a bit farther back into its slot using a short dowel to protect your fingers, then watch to see that the blade freely springs back against the jig when the pressure is released (see the photo above). When you're satisfied that the knife is positioned properly, snug up the two end bolts to hold the blade in place, then remove the jig and progressively tighten the remaining bolts.

Some planers come with a jig that positions only one end of the knife at a time. Using this type of jig is slower and more prone to error since moving the blade against the jig at one end also slightly shifts the opposite end. You will be forced to go back and forth several times, loosening, resetting, and then tightening the ends with smaller and smaller adjustments until both ends of the blade are at the correct height.

To minimize the degree of shifting and speed up the process when using a jig that sets one end of the blade at a time, always place the jig as close as possible to the outermost lock bolt as you set each end of the knife. Often, you can convert a simple jig into one that will set the blade height at both ends by purchasing a second jig from the machine's manufacturer and attaching the two jigs together with a dowel handle that's about three-quarters of the blade length.

INSTALLING JACKSCREW-SET KNIVES

On some planers, the blade height is adjusted by jackscrews that lift the blade into position and is checked using a dial indicator mounted in a base supplied with the machine. The basic procedure is simple. To begin, zero the indicator against the outer surface of the head on the rounded area between two of the knives (see the photo below). Be careful to make sure all of the contact points are free of dust. Once the indicator is zeroed, carefully set it aside where it won't be jarred and lose its setting.

Place a new knife in its slot and adjust the lock bolts to leave the knife with a small amount of free play. Place the indicator, which was zeroed against the head, over the knife, close to one of the jackscrews (see the photo on p. 110), then shift the indicator so it reads the high point of the edge. The knife should be below the height specified in the manual. If not, back off the adjusting screws and tap the blade down to seat it in the notches of the jackscrew's housing. Next, set the two lock screws nearest the jacks to exert a slight drag on the knife while allowing it to move upward as you adjust the jackscrews. The slight drag improves accuracy by eliminating free play of the knife in the slot and by keeping the back of the knife snug against the jackscrews.

Using a jackscrew, raise one end of the knife to a point a few thousandths of an inch below the specified height. Move the indicator and repeat this procedure at the opposite end of the knife, raising the knife to just below the specified height.

The first step in setting knife height is to zero the indicator against the cutterhead between knives.

Using the jackscrews, adjust the knife height while measuring the blade's exposure with the indicator.

If the machine's manual is missing and you don't have a blade-height specification, bring the knife up until the full width of the bevel on the back edge is exposed and note the indicator's reading. Now add a couple of thousandths of an inch to the reading and use this number as the indicator reading to set all of the knives in the head. If you have to set the blade height using this method, for safety always turn the cutterhead through one complete revolution by hand to be absolutely sure the knives won't strike some part of the housing.

Once both ends of the knife are just shy of full height, repeat the process a second time to take the blade up to its exact height. Be careful not to overshoot the mark. If you do go over, back off the jackscrews slightly, tap the blade down, and try again. You should reach the final setting while raising the blade out of the slot with the jackscrews, never while tapping the knife to lower it. Once you've set the correct height, gradually tighten the lock screws following the procedure outlined earlier in this section.

Installing disposable knives After checking blade-locking components, install disposable knives following the instructions that came with the machine. In most cases, there is little that can go wrong. Just make sure the knife and its mounting surfaces are clean, flat, and burr free. The thin blades can be easily distorted and may be prevented from seating and aligning properly.

To check cutterhead alignment, start by adjusting the table to bring the indicator's hand to the vertical position after it has made a couple of turns.

Move the indicator to one side of the cutterhead and adjust it so it reads zero.

Move the indicator to the opposite end of the cutterhead and take another reading. Here the head is out of parallel by 0.005 in.

Aligning the Table with the Cutterhead

To plane a board to an even thickness, the cutterhead must be parallel to the surface of the table. It's easy to check this with a dial indicator on a simple base. If the table and head aren't parallel, correcting the alignment is relatively simple, although the exact procedure will depend on the design of the planer.

To check for parallel, make the indicator base described in the sidebar on p. 97, and attach the indicator with its stem pointing upward. A large, flat tip will work better than a small one. Lower the table to clear the indicator assembly, and rotate the cutterhead until a smooth portion of the head, midway between two knives, is facing the table. Place the indicator in the middle of the table under the approximate center of the cutterhead, and adjust the table height to make contact between the tip and the rounded surface of the head.

Continue to move the table to rotate the indicator's hand a couple of turns. Slide the indicator from front to back to locate the lowest point of the head, then with the indicator positioned, raise or lower the table slightly to bring the indicator's hand to the 12 o'clock position and lock the table or head in place (see the photo at left above).

Now place the indicator under one side of the head, moving it from front to back to locate the low point and zero the dial (see the center

On most portable planers, there is a geared shaft or chain drive connecting the left- and right-side jackscrews. Unhooking the linkage allows you to raise one side of the head to make it parallel with the table.

photo on p. 111). Move the indicator to the opposite side of the table, position it under the lowest point on the head, and take a reading. If the dial doesn't still read zero, the cutterhead and the table aren't parallel (see the photo at right on p. 111).

Correcting an out-of-parallel table is simple, at least in theory. You turn the jackscrews, or the nuts they go into, on just one side of the table to bring it into line. If you have the owner's manual, it may tell you how to make the adjustment. If you don't have specific information, start by looking under the table or the powerhead, depending on which moves, to see if there is a threaded bushing that the jackscrew goes into. If you can see a bushing that is hex shaped or has holes around its perimeter for a bar to turn it, loosen the bushing's clamps or set screws and turn the bushing to align the table and cutterhead.

If you can't find a bushing to adjust, temporarily disconnect the chain, gear drive, or cogged belt that connects the left- and right-side jackscrews, and turn one of the screws to realign the table (see the photo above). If you have a machine with four jackscrews, turn both of the screws on one side equally to adjust the alignment. After making an adjustment, recheck with the dial indicator, then lock up the set screws or reconnect the gear train.

Infeed and Outfeed Roller Adjustments

Roller heights are set relative to the knives in the cutterhead, so the knives should be sharp and properly adjusted before making any adjustments to the rollers.

One Last Check

As a final check for parallel, you can rotate the head to place a knife at the bottom dead center position and measure between the table and the knife for equal dimensions on the left and right sides of the table. If the knife isn't parallel to the table, double-check that the blade is properly installed before making any adjustments to the jackscrews.

Set infeed and outfeed roller height by adjusting screws at each end of the rollers after taking a measurement with a dial indicator.

Begin adjusting the infeed roller by manually turning the cutterhead until one of its blades is at the bottom of its arc over the table. Set up the dial indicator on its base with a wide, flat tip on the plunger. Place the indicator under the middle of the cutterhead, and raise the table until the indicator's tip is touching the blade and the hand has made a couple of turns. Gently rock the cutterhead back and forth and shift the indicator's base until you are sure the blade is at the lowest point of its arc and the indicator is centered directly below it. With the head properly positioned, raise or lower the table slightly to bring the indicator's hand up to the 12 o'clock position. If the height-adjustment mechanism has a lock, set it now. Finally, zero the indicator and note the value on the tenths dial.

With the indicator zeroed, slide it underneath the left end of the infeed roller. If the roller is corrugated, rotate it to place a raised segment facing down toward the table. Take a reading, checking that you are still in the same tenths range. The infeed roller should be set slightly closer to the table than the arc of the knives on the cutterhead. If you don't have specifications, try setting the infeed roller 0.030 in. closer than the knives for a solid roller and 0.060 in. closer for a segmented roller. Once the roller height is properly set, the hand of the indicator will rotate one-third to a little more than one-half of a revolution clockwise when you move from under the knives to under the roller.

Setting roller height is easy on most planers. An adjustment bolt with a locknut is located under the roller's bearing housings at either end of the roller (see the photo above). To reset the height, loosen the bolt and thread it in or out until you have the reading you want on the dial indicator, then tighten the locknut.

Segmented Rollers

Taking a reading against a segmented infeed roller can be difficult. Segments float on springs, and they are not concentric with the inner shaft that drives them. Try disconnecting the roller's drive chain so the roller rotates freely, then tap the segment you want to read until it runs nearly concentrically. Using the dial indicator, find the segment's high and low points, which will be 180 degrees apart. Midway between these two points, you can get an accurate reading of the roller's height. Once you've located and marked these midpoints on segments on either end of the roller, you can set its height.

On some planers, the spring pressure on the feed rollers can be adjusted. Stock that feeds poorly or skews in the machine is a sign the adjustment may be necessary.

Using the indicator and its base, set the height of the chip breaker so it is slightly lower than the arc of the knives.

The outfeed roller is also set closer to the table than the cutterhead; if you don't have specifications, set it 0.030 in. closer than the knives. The procedure for checking and setting the height of the outfeed roller is identical to the method just outlined for setting infeed roller height.

SETTING ROLLER-SPRING PRESSURE

The infeed and outfeed rollers are pressed tightly against the board being planed by springs above the bearing housings at the ends of each roller. Spring pressure is generally not adjusted on a routine tune-up. On many machines, the springs can't be adjusted.

On planers that do allow you to adjust spring pressure, the setting would only be changed if there are feeding problems and all other possible causes have been eliminated. Typically, most feed problems are caused by excess friction between the table and the board, dull knives, or the stock binding against the chip breaker or pressure bar. However, there are some situations where spring pressure should be adjusted. If stock consistently skews when the leading end of the board first passes under the infeed roller (and the bed roller is set correctly), increase the pressure on the side that isn't pulling as well (see the photo at left on the facing page). On some machines the tension on the end of the roller driven by the chain drive needs to be slightly increased to compensate for torque and pull of the chain.

If the rollers slip excessively and there are no friction or binding problems, increase the spring pressure at both ends of the roller. In time, springs can weaken, and it may be necessary to replace the springs if tightening them doesn't correct slippage.

If the feed roller is corrugated and you primarily run softer woods, such as pine and poplar, you can decrease spring pressure on the infeed roller so the roller is less likely to leave an imprint on the stock.

CHIP-BREAKER ADJUSTMENT

Like the rollers, the chip breaker extends slightly below the arc of the knives. If you don't have a specific setting from a service manual for your machine, set the bar 0.030 in. below the level of the knives. On some planers, the chip breaker rides up and down with the infeed roller assembly, so the breaker should be adjusted only after the infeed roller height has been set. For the same reason, the chip breaker's height should be checked after any adjustments are made to the infeed roller height.

In setting the chip breaker, follow the same procedure used to set the roller heights. That is, use a dial indicator zeroed on a blade in the cutterhead (see the photo at right on the facing page). Depending on the design of your machine, the two adjusting bolts, one at each end of the breaker assembly, can be reached from either above or below the head assembly.

PRESSURE-BAR ADJUSTMENT

The pressure bar holds the stock down as it leaves the cutterhead. If the bar is set too high, the planer will snipe the ends of the board. Unlike the chip breaker and feed rollers, the pressure bar is not spring-loaded; it won't ride up to clear the stock if it is set too low. To prevent jamming, it must be set precisely, ideally no more than 0.001 in. to 0.002 in. higher than the arc of the knives.

Use the same procedure you followed for setting the chip breaker and feed rollers. Zero the indicator on the knives at bottom dead center, then take a reading with the indicator under the bar. To set the pressure bar accurately, adjust its position immediately after you have installed new knives in the machine. Set the height 0.002 in. higher than the knives (see the top photo on p. 116).

The height of the pressure bar on the outfeed side of the planer is critical to good planing. Set too high, it will allow sniping; too low, jamming.

To check the height of bed rollers, mount the indicator on its base pointing downward, and zero the dial against the table's surface.

BED-ROLLER ADJUSTMENT

To reduce friction that might stall a board, large planers have unpowered rollers recessed into the bed directly under the infeed and outfeed rollers. Properly set, they stand a few thousandths of an inch above the level of the table. Some planers have control levers that allow you to adjust the bed roller's height to obtain the smoothest feed for the type of wood being planed. On machines with fixed-height bed rollers that can only be adjusted by undoing bolts and locknuts, you may have to experiment to find the best height for stock you typically plane.

Use the same procedure for adjusting both the infeed and outfeed bed rollers, which should be adjusted to the same height. Begin by attaching the dial indicator to its base with the indicator's plunger projecting downward through the ½-in. hole in the base, being sure to use a flat, wide tip on the end of the plunger. Set the base on a flat surface and adjust the three leveling screws by approximately equal amounts to bring the indicator's hand up to the 12 o'clock position after the hand has made a couple of revolutions from its unloaded position (see the bottom photo on the facing page).

Move the indicator and base onto the planer's table, and zero the dial with the tip bearing against the table. Next, place the indicator over the end of the infeed bed roller, moving the base from front to back to position it over the highest point on the roller's surface. Rotate the roller through one complete revolution while keeping an eye on the indicator. The hand should show a variation of no more than a couple of thousandths of an inch. If the variation is greater and it is not caused by a burr or pitch on the roller's surface, then the roller is out of round, bent, or has a bad bearing.

Repeat this check over the other end of the roller and at its center. If you measure the greatest variation at the roller's midpoint, the roller has probably been bent. You'll need to replace or repair a wobbly roller because it would cause feeding problems, usually in the form of an intermittent binding every few inches as the board is fed through the machine.

If the roller is running true, check that the indicator is still zeroed when its tip is on the table's surface. If the roller's height is adjusted by a lever, set the control to the zero or lowest position. Place the indicator back over one end of the roller, centering it over the high point, then take a reading and check the other end of the roller.

On a lever-adjusted roller, the roller, at its lowest setting, should project above the table by 0.006 in. to 0.008 in. at each end and the two measurements should be the same. On fixed-height rollers, try a beginning height of around 0.010 in. for planing hardwoods and 0.015 in. for softwoods. The best height can only be determined by experiment. You want the minimum height needed for smooth feeding because any additional projection of the rollers can cause binding against the pressure bar and

The bed-roller bearings on this planer are mounted slightly off center in their housings. Roller height is adjusted by rotating the housings.

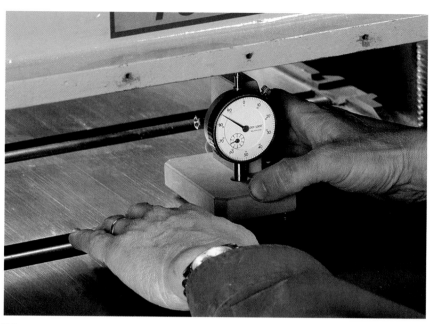

When used on the bed rollers, the indicator can check initially for a bent roller or worn bearings.

possibly a reverse snipe, in which there is a higher step on the trailing end of the stock.

The height of each end of the roller is typically adjusted by a screw and locknut reached from underneath the table. On some planers, the bed-roller bearing housings are eccentric and the roller height is set by loosening a set screw and rotating the housing in the planer's frame. Lever-adjusted rollers can have either a screw-and-nut arrangement or the linkage may need to be uncoupled so that each end of the roller can be moved independently with its jackscrew.

After setting the infeed bed roller, repeat the same procedure for the outfeed bed roller, setting it to the same height as the infeed side and even in height at both ends.

RUNNING A TEST BOARD

Once you've cleaned up the planer, installed new knives, and adjusted everything, it's time to run a test board (see the photo on the facing page). This is primarily to check the pressure-bar setting. As explained in "Troubleshooting" on p. 94, poorly prepared stock or feeding the board improperly can cause sniping that is not the fault of the machine.

The final step in tuning a planer is to run a test board. Once the machine is tuned, it should hold its settings for a long time.

To test, feed the board while taking a light cut and watch for jamming under the pressure bar. If the stock jams, raise the bar a few thousandths of an inch at a time until the jamming stops. The more likely case is the board will have snipe on either end and the pressure bar can be dropped a few thousandths—as long as it doesn't lead to jamming. To prevent jamming later, you may have to accept a tiny amount of snipe, 0.001 in., with new knives. It will disappear as the knives wear in.

Finally, check the thickness of the planed board on the long edges. If the board consistently reads thicker on one edge than the other, check and readjust the table-to-head alignment. Once the machine is planing well, just keep it clean and well lubricated and its knives sharp. A full tune-up shouldn't be needed again for a long time.

The Drill Press

A drill press is a relatively simple machine that should enjoy a long life. Even older machines like this one should operate smoothly and accurately with only a few periodic adjustments.

A drill press was an important part of my first attempts at professional woodworking. I discovered that a drill press made joinery with screws, bolts, and dowels fast and accurate, and efficiency is what paid the rent. I still make much of my furniture this way, hiding the hardware inside the carcase or concealing it with plugs and moldings.

The drill press is one of the oldest kinds of machine tool. Early versions, in fact, were powered by hand. Circular saws, bandsaws, jointers, and planers all require more power than people can generate, so they only became possible with the development of water, steam, and electric power.

Keeping a drill press in good working order is not difficult. There are many fewer adjustments to make than there are on a more complicated machine such as a jointer or thickness planer.

Anatomy

As is the case with many power tools, the design of the drill press settled into its standard form more than a century ago and has changed little since. The backbone of the tool is the post or column, a thick-walled steel tube 2 in. to 4 in. in diameter supported by a base casting. At the top of the column is a head assembly that houses the drive and control components. On a modern drill press, the motor is mounted on the back of the head assembly, but its position can be shifted to adjust belt tension.

Power is transmitted to the bit by a V-belt running on stepped pulleys (see the top photo on p. 122). The pulleys allow you to adjust the speed of

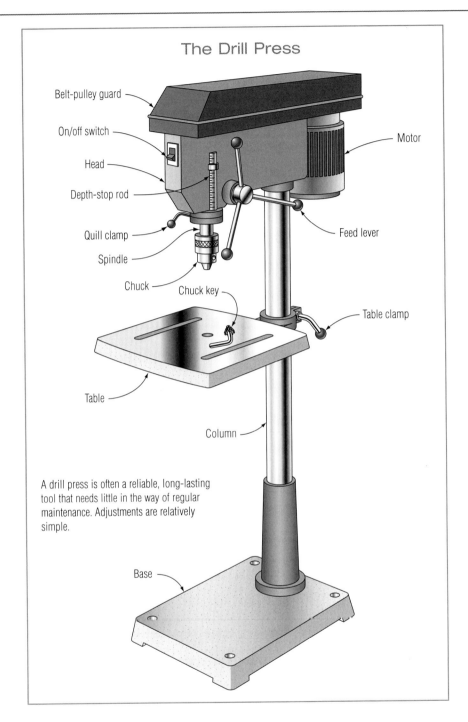

The Drill Press

Belt-pulley guard

On/off switch

Head

Depth-stop rod

Quill clamp

Spindle

Chuck

Chuck key

Motor

Feed lever

Table clamp

Table

Column

A drill press is often a reliable, long-lasting tool that needs little in the way of regular maintenance. Adjustments are relatively simple.

Base

the bit. Generally, the larger the bit, the slower it should turn (see the chart on p. 123). Some drill presses offer a greater speed range by using two belts and a third step pulley midway between the motor and spindle, but shifting a belt to change speeds is time consuming. Having a variable-

You can change drilling speed by shifting the position of the belt so it engages different steps on its pulleys. This one is set up for its lowest speed.

To move the drill bit up and down, a gear rack machined into the surface of the quill engages a gear turned by the feed lever.

speed drive is a real advantage if you use the tool with a wide range of bit sizes. This feature, which works something like an automatic transmission in a car, is more common on heavier, industrial-grade tools.

The front pulley drives a spindle on which the drill chuck is mounted. The spindle runs in ball bearings housed in a movable, precisely machined tube called a quill. To move the spindle up or down, a gear on the feed-lever shaft engages a rack machined into the outside of the quill (see the photo at left). When the gear is turned by rotating the feed lever, it moves the quill, spindle, and chuck as a unit. To transmit power to the spindle, the bore of the pulley is lined with ridges called splines that engage matching splines on the spindle shaft.

The opposite end of the feed shaft is attached to a flat coil spring enclosed in a round housing. The spring reverses the motion of the gear when the feed lever is released, returning the quill to its retracted position (see the photo at left on the facing page). To control the depth of drilling, a collar with a threaded depth stop is normally clamped to the bottom of the quill just above the chuck (see the photo at right on the facing page).

The last component of a drill press is the table that rides up and down on the column. On all but the smallest machines, a gear-and-rack mechanism for raising and lowering the table is built into the collar attaching the table to the column. On many drill presses, the table is mounted on a pivot, allowing it to tilt for drilling angled holes. If you rotate the table to a vertical position, you can drill holes in the ends of a workpiece (see the photo at left on p. 124). Many tilting tables have a removable pin that, in theory at least, locks the table so it is square to the bit (see the photo at right on p. 124).

DRILL PRESS SUGGESTED SPEEDS (IN RPM)

Diameter	Softwoods	Hardwoods	Aluminum	Brass	Cast Iron	Mild Steel
¹⁄₁₆ in.	4,700	4,700	4,700	4,700	4,700	2,400
⅛ in.	4,700	4,700	4,700	4,700	2,400	1,250
³⁄₁₆ in.	4,700	2,400	4,700	2,400	2,400	1,250
¼ in.	2,400	2,400	4,700	2,400	1,250	700
⁵⁄₁₆ in.	2,400	1,250	2,400	1,250	1,250	700
⅜ in.	2,400	1,250	2,400	1,250	700	700
⁷⁄₁₆ in.	2,400	1,250	1,250	1,250	700	
½ in.	1,250	1,250	1,250	700	700	
⅝ in.	1,250	700	700			
¾ in.	1,250	700	700			
⅞ in.	1,250	700				
1 in.	700	700				
1¼ in.	700	700				
1½ in.	700	700				
2 in.	700	700				

Behind the cover (left) is a large spring that pulls the quill up when the feed lever is released. Next to the housing is a lever that can lock the quill in position.

Precise control over the depth of a drilled hole is one of the advantages of using a drill press. A nut on the threaded depth gauge stops the quill's descent, and the large O-ring around the bottom of the quill cushions it when it is fully retracted.

Tightening Up the Nuts

On most drill presses, the depth-stop rod has two knurled nuts on it that are supposed to lock together to hold the depth setting. These nuts often loosen up, allowing the drill bit to go deeper than you planned and quite probably ruining the workpiece. To prevent this, place a thick O-ring that fits loosely over the rod between the two nuts. When the nuts are tightened against the O-ring, they will stay put.

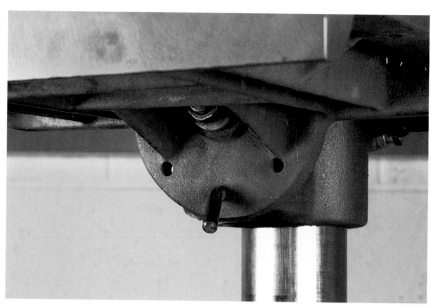

A pin set into the underside of the table-tilting assembly makes it easy to lock the table so it is square to the bit.

By swinging a tilt table to its vertical position, you can drill the ends of a workpiece easily.

Troubleshooting

Drill presses suffer from relatively few problems, and most of them are easy to fix. The most common of them is a table that is not square to the drill bit. Depending on the design of the machine, it can be difficult to adjust the position of the original table. One solution is to add an auxiliary table that you can shim into position easily.

Drill chucks rarely malfunction. On poorly made or abused machines, the chuck may be worn or mounted crookedly on the spindle or the spindle could be bent. These problems cause the drill bit to wobble, leading to over-sized and out-of-position holes. You can measure runout with a dial indicator, but excessive wobble is fairly obvious when the machine is running. Runout should not exceed 0.010 in. as measured on the bit 2 in. below the chuck. Accurately diagnosing and repairing a drill with runout problems can be difficult. If you are having persistent problems with wobbling drill bits and the machine is worth the expense, take it to a machinery-repair shop. This repair is beyond the scope of most home shops.

Bearing problems also are rare, primarily because the bearings are well protected from dust and run at only moderate speeds. In contrast, the motors on drill presses have slightly more problems than motors on most other woodworking tools. The reason is that they are used in a vertical position, which most standard motors are not designed to do. As the motor's lower thrust bearing wears under the excess load, the internal starter switch will malfunction. As a result, the motor may fail to start or only run slowly and then overheat after a few seconds of loud humming.

The fix for this is usually simple: Remove the lower bell housing from the motor and add a shim washer or two on the end of the shaft to raise it slightly, compensating for the worn bearing.

Checking the Machine's Basic Setup

Begin with a look at the machine's bearings, belts, and pulleys. Remove the drive belt, then turn the spindle by hand. If bearings are in good condition, there will be no noise, roughness, or sticking. Next, check the bearings in the motor by spinning the shaft by hand.

Examine the step pulleys on both the motor and spindle shafts: They should turn with little or no visible wobble. On smaller machines, the pulleys are made from die-cast metal and can wear out. If you need to replace them, you'll probably have to go to the machine's manufacturer for parts because the front pulley may not be a standard bore. Make sure the pulley's set screws are tight; check for screws in the bottom of every groove. Some pulleys will have two screws (see the photo below).

Check the belt for stiffness, wear, and cracking before placing it back on the pulleys. You can greatly improve the performance of a drill press by replacing the standard belt with a PowerTwist linked belt made by Fenner Drives (see the top photo on p. 126). An added advantage of a linked belt is that it has a little stretch to it, meaning you can change drill speeds by slipping the belt over the rim of the pulleys without having to undo the belt-tensioning device, which, on many machines, is poorly designed and difficult to adjust.

When tightening a pulley, look in the bottom of each groove. There may be more than one set screw.

Replacing the stock V-belt with a linked belt reduces vibration and makes it easier to change belt positions.

Often poorly designed, the tensioning mechanism will work better after removing paint and burrs and lubricating the sliding parts. You can use a short board for leverage between the post and motor when tensioning the belt.

In addition to increasing vibration, a misaligned belt causes excess wear on the pulleys and shortens belt life. Here, the pulley should be lowered.

Next, stand behind the drill press to check the position of the motor. Correctly installed, the motor will be vertical, not canted to one side. Repeat this same check from the side. If the motor is misaligned, loosen its mounting bolts and straighten it, using shim washers if necessary. At the same time, check the belt-tensioning mechanism, which should work smoothly. You should remove any burrs or paint that cause the parts to bind and lubricate all the moving parts in the assembly with oil or grease (see the photo at left on the facing page).

Reinstall the belt and look at it from the side. Properly aligned, the belt leaves the pulleys squarely, not angling up or down as it leaves the grooves (see the bottom photo on the facing page). If the alignment is off, loosen the motor pulley and shift it higher or lower on the shaft to get the belt running straight. When tightened correctly, the belt should deflect about ½ in. when you press against it firmly with a finger.

Check that the clamps or set screws securing the column to the base are tight, and make sure the head is directly over the base. If the head is in line, be sure that its clamps or set screws are tight. If the head is misaligned, unclamp and pivot it to line up with the base. But be careful: Loosening these bolts will allow the head to drop suddenly, smashing the chuck, the table, and quite possibly your fingers. To prevent this, tighten a spiral hose clamp around the column just under the head before loosening any of the bolts (see the photo at left below).

Large machines often have a safety collar clamped around the post for this same purpose. Make sure the collar is up against the head and its bolts are tight. As an added precaution when shifting the head, place a board on

By preventing the head from sliding down the post, a hose clamp makes it safe to loosen head bolts and adjust head-to-base alignment. It can be left in place permanently.

Penetrating oil cleans and lubricates the chuck. Be sure to leave newspapers under the chuck overnight to catch the last few drops of oil.

the table and bring the table up close to the chuck before undoing the bolts. If something lets go, the board will prevent the chuck from slamming into the cast-iron table.

The column should be free of burrs and rust so the table moves up and down smoothly. Clean the column with an abrasive pad, using a file to remove burrs, and then apply some wax. If the table has a track and gear to adjust its height, clean up the track and give it a light coat of a stick lubricant. Make sure the clamps or screws holding the track to the column are tight and the track is vertical. Finally, apply a couple of drops of oil to the shaft of the table's crank handle and to the shaft of the feed lever.

If the table can be tilted, remove the pivot bolt and examine the circular bearing surfaces of the castings. They should be smooth and free of paint. Clean and file them as needed and reassemble after greasing the bearing faces and the pivot bolt threads.

Lubricate the drill chuck by opening it up wide and spraying penetrating oil inside using an extension nozzle (see the photo at right on p. 127). To catch drips, keep a rag handy or use a layer of newspaper and a shallow pan. The oil may flush out dirt and wood chips. As you spray the oil, run the chuck from the closed to the wide-open position a few times to work the oil into all of the parts of the mechanism. If the chuck feels very gritty as you turn it, clean it out first with spray automotive brake cleaner followed by a spray of oil. Lastly, use a rag or a paper towel wrapped around a dowel to remove the excess oil from the chuck's jaws.

Tuning Up a Drill Press

Once you have your drill press working smoothly, there's not much left to do. The only major adjustment is to square up the table so it is perpendicular to a chucked drill bit. You also may need to adjust the feed-lever return spring.

START WITH A NEW AUXILIARY TABLE

The small metal tables standard on most drill presses are designed for working on metal parts. They are too small to support the large workpieces typical in cabinetmaking, and the ribbed underside of the table makes it difficult to clamp either a fence or stock. Bolting a larger table, made from ¾-in. MDF or plywood, on top of the original table will solve these problems and make it easy to square up the table to the axis of the drill chuck.

An added advantage of adding the auxiliary table is that MDF or plywood won't dent your stock as readily, and dropping a bit or accidentally drilling into the table's surface won't damage your drills. The top, if damaged, can be easily replaced. Even so, I use scrap wood under the workpiece whenever I drill through. It prolongs the life of the table and minimizes chipping and tearout on the back of the stock.

A large auxiliary table is better suited to woodworking than the small, metal table that comes with a drill press.

As shown in the photo above, the auxiliary table is very simple, just a rectangle that's 2 in. wider than the metal table at the front and back and 4 in. to 6 in. longer than the original table at the sides. The table could be made larger, especially in length, but MDF would need reinforcement to prevent it from sagging under its own weight. If the back edge of the origi-

Simplifying an Auxiliary Table

To make an easy auxiliary table, just drill a hole in the new table to match the one in the metal table below. You also could cut a square hole in the center of the top and use inserts that you could easily change if damaged. To avoid bruised ribs when working around the drill press, round or miter off the table's sharp corners.

The broad heads of bolts made for knockdown furniture make them ideal for attaching an auxiliary table to a drill press.

nal table is close to the column, notch the new table to go around the column to get a full 2 in. of overhang along the back edge.

Bolt the auxiliary table onto the original table using ¼-in. flat-head bolts and nylon-insert locknuts. If you want to be more elegant, use the wide-head bolts made for assembling knockdown furniture (see the photo above). Cast iron, the material the original table is probably made from, will drill very easily with a sharp bit. When locating the holes for the attachment bolts, avoid the ribs cast into the underside of the metal table and be sure to leave sufficient clearance for placing a washer on the bolt. Although typically there are slots milled into the metal tops, they are often in the wrong place for use in mounting an auxiliary table and they are too wide.

SQUARING UP THE TABLE

To square up the table, you need to install a reliably straight rod in the drill's chuck to use as a reference. I use a ⅜-in.-dia. extra-length bit that costs about $10. Extra-length bits are about 1 ft. long with the flutes running only part of the way up the shank, leaving the upper half smooth.

An extra-long drill bit provides an accurate reference when checking whether the table is square to the centerline of the drill chuck.

Use All the Keyholes

To properly tighten a drill chuck, use the key in all three of the keyholes. Go around once to snug up the bit, then go around a second time, tightening with the key in all three holes to fully lock the chuck. Using this technique, the bit will be held tighter and straighter in the jaws, avoiding slippage and wear on the chuck.

If there isn't enough room under the chuck to use the square in its normal orientation, lay it on its side and use the base against the bit.

First, install the long bit in the chuck, then raise the table until only the smooth upper half of the drill's shank is above the table's surface. Once the table is in position, be sure to tighten the clamp to the column. Your measurements will be thrown off by any looseness in the fit between the table's collar and the post.

The table needs to be square in two directions, from side to side and from front to back. Start by placing a square against the side of the bit. If the table is out of square and has a tilt function but no lock pin for the 90-degree position, the adjustment is easy. Just loosen the pivot bolt and square up the table to the bit. If your drill press doesn't have a tilting table or if it locks at the 90-degree position, you'll still be able to square it up by adding shims between the metal and auxiliary tables.

Cut the shims from aluminum flashing for large adjustments or from a soda can (which is thinner) for smaller changes. Ideally, the shims should be in the form of washers that slip over the mounting bolts, but for now, just cut small tabs of shim stock and slide them in next to the bolts (see the top photo on the facing page). To prevent twisting the table, always install the shims in pairs of the same thickness, one by each bolt at the front and back on either the left or right side of the table. After slipping in the shims, retighten the bolts before checking to see if they have corrected the table's tilt.

Once you have fixed the table's tilt to the side, place the square against the front edge of the drill's shank to check if the table tilts from front to back. The shimming procedure is the same, except that the pair of shims is installed next to either both of the front or both of the back bolts. If you already have shims in place for correcting side tilt, add the new shims directly on top of them.

When the table is square in both directions, make permanent shims that fit around the bolts and better support the wooden top. Unless you have a punch for making the center holes, making shim washers is difficult. A simple but adequate shim for this application is one in the shape of a fat L cut from a piece of sheet metal about 2 in. square.

The easiest way to install the shims is to remove the MDF table and position the permanent shims on the metal table with a small dot of Super Glue. The legs of the L bracket the bolt hole as shown in the photo on p. 134. If you ever have to change the shims, you can remove them by heating the shim briefly using a torch. The heat causes the glue to break down.

Temporary shims square the table to the bit and can be replaced by permanent ones that fit snugly around bolts that secure the auxiliary table.

Annealing the Spring

The hard, tempered steel used to make the return spring for a drill press is brittle. Occasionally, it breaks—typically at the bend where the spring is attached to the shaft or the housing. You can reuse the spring and be back to work in a few minutes by bending a new hook in the end. The spring steel is too stiff to bend without annealing it first to make it softer and less brittle. To anneal the spring, clamp it in a metal vise with a few inches of the broken end showing above the jaws. Heat the exposed end to red hot for a few seconds using a propane torch, then allow the spring to cool slowly so it will be soft enough to bend easily.

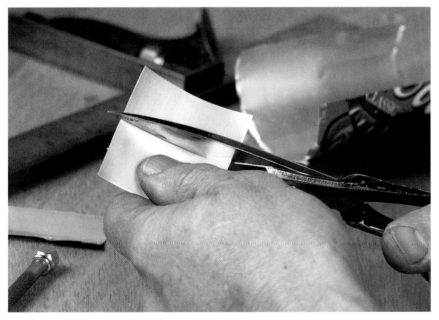

L-shaped shims are easy to make from aluminum flashing or soda cans.

SETTING SPRING TENSION

Setting the tension on the feed-lever return spring is the only other adjustment to make. If you regularly have problems backing bits out of their holes, you should increase the tension. If you have the opposite problem—that is, the quill pulls back too forcefully—you should reduce the pull of the spring. Attached to one end of the feed-lever shaft, the spring, which is a flat coil of the same type used in a wind-up clock, is contained in a round housing on the side of the head assembly. One end of the spring attaches to the shaft, while the other end is clipped onto the housing.

To adjust the spring tension, loosen the housing slightly so you can rotate it and wind or unwind the spring. On some drills, the housing is held in place by a center bolt. When the bolt is backed off a couple of turns, it allows a notch in the rim of the housing to lift over a pin in the side of the head casting. Once over the pin, the housing can be turned and reengaged in a different notch. Alternately, the spring housing can be held by a set screw or several bolts catching a flange on the housing's rim. With

A dot of Super Glue keeps the shims from shifting and keeps stacked shims lined up. If the shims need to be removed, the heat from a propane torch will break the glue bond.

Keep a firm grip on the spring housing as you loosen it. The spring is under tension and will try to turn the housing counterclockwise.

this style of attachment, the set screw or bolts simply have to be loosened slightly to allow the housing to turn and change the spring tension.

Be aware that the spring housing is under tension and will want to turn forcefully the moment it is free. When you start to loosen the bolts, have a good grip on the housing or it can unwind very abruptly (see the photo above). You only have to loosen the bolts just enough to turn the housing; don't pull the housing off completely or the spring will pop free and unwind with considerable violence. To be safe, always protect your eyes with safety glasses whenever you adjust the spring tension or have to remove the housing. Once you've tuned up your drill press, it should give you several years of service before it needs to be checked and lubricated again.

Let the Spring Unwind

If you ever have to remove the spring and its cover completely—to service the feed gear or remove the quill assembly, for example—allow the spring to unwind completely before removing the housing. Once the spring has fully unwound in its housing, you can remove the whole assembly safely. Be sure to lift the housing off carefully so the spring is not pulled from the housing, and wear safety glasses to protect your eyes just in case the spring does come loose.

6

The Bandsaw

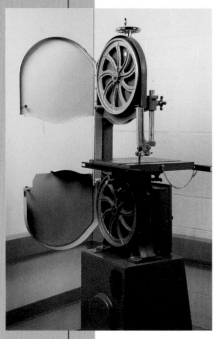

Useful for much more than sawing curves, a bandsaw is unsurpassed for ripping thick or warped stock and can be effectively used for cutting dovetails and tenons.

My first encounter with a bandsaw was disappointing. The saw was a small Sears Craftsman made in the 1950s that arrived in my shop as part of a package deal on several used power tools. I could never get the machine to cut well, and blades snapped regularly. I rarely used it and after a few years I sold it. I was convinced that bandsaws were cranky tools of limited use to anyone not making jigsaw puzzles. Since then I've learned how to set up and use a bandsaw, and I'm sorry I sold that little machine. It was very well made, and with the right blade and a tune-up, it would have been a real asset.

A bandsaw is much more than a specialty tool for cutting curves. It can cut tapers without the use of jigs, and it is an excellent tool for ripping lumber, rivaling a table saw. It offers great depth of cut and zero chance of kickback, no matter how gnarly the stock.

For joinery, a bandsaw makes short work of roughing out dovetails and sawing tenons. In many cabinet shops, the saw's main use is for resawing, cutting a thick board into several thinner ones. Resawing makes efficient use of expensive or highly figured wood, and it makes it easy to match grain. In fact, a careful furniture maker can sometimes cut all of the most visible elements in a piece of furniture from a single board.

Bandsaws are complex machines. They have more moving parts and adjustments than any other three power tools combined, and they are unforgiving of misalignments and poor maintenance. Despite their complexity, however, properly setting up and tuning a bandsaw is not difficult, just time consuming. Once the saw is working well, its versatility will be an ample reward.

An ability to resaw thick stock greatly expands furniture-making options.

Anatomy

The thin ribbon of steel that forms the blade of a bandsaw must be pulled very taut to cut properly. The tension on a ¾-in. blade can exert almost 1,000 lb. of stress on the frame of a bandsaw, so to resist this force, the frame must be sturdy. In large saws, the frame is made of cast iron, whereas smaller benchtop models are often made from die-cast aluminum. Several companies build their bandsaw frames from steel plate.

The Bandsaw

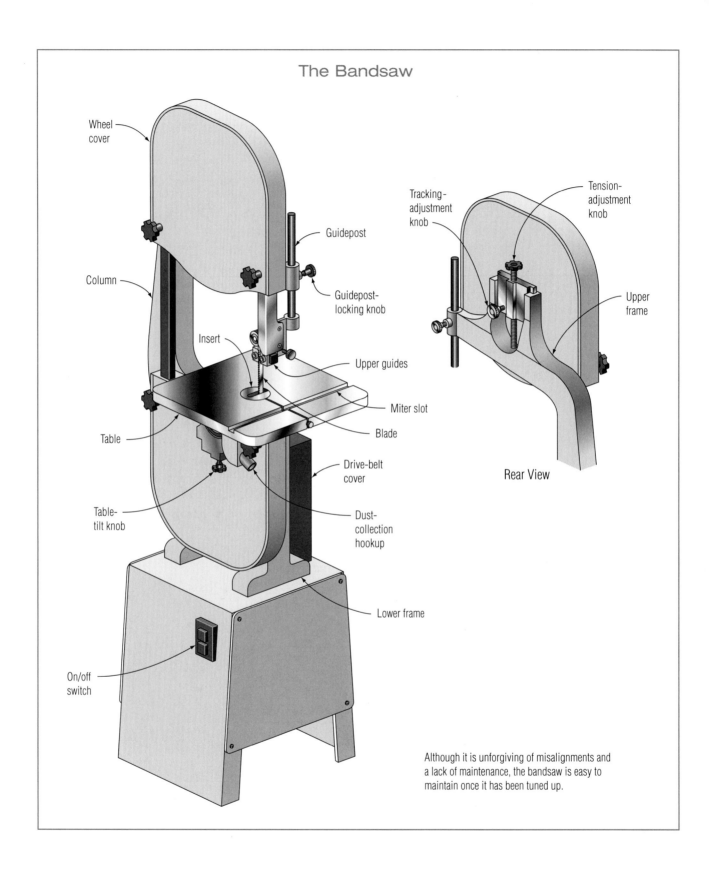

Wheel cover

Column

Guidepost

Guidepost-locking knob

Insert

Upper guides

Miter slot

Table

Blade

Table-tilt knob

Drive-belt cover

Dust-collection hookup

Lower frame

On/off switch

Tracking-adjustment knob

Tension-adjustment knob

Upper frame

Rear View

Although it is unforgiving of misalignments and a lack of maintenance, the bandsaw is easy to maintain once it has been tuned up.

The popular, and often imitated, Delta 14-in. bandsaw has a two-piece cast-iron frame that bolts together at the bottom of the back post. You can increase its cutting capacity by installing a riser block between the upper and lower castings.

THE WHEELS

At the heart of a bandsaw are the wheels that the blade travels around. Made from steel plate or cast aluminum or iron, they can run from a petite 6 in. to a behemoth 10 ft. or more in diameter. Wheels measuring 14 in. to 24 in. are typically seen in cabinet shops. On better-quality machines, the wheels are carefully machined and balanced to eliminate vibration.

The rims of the wheels are covered by a rubber tire a fraction of an inch thick. On most machines, the tire has a slight curve in its surface, called a crown, making it higher in the middle (see the illustration at right). A crowned tire helps to pull the blade to the center.

Most saws use a pair of wheels, but occasionally you may find a three-wheeled machine. A third wheel increases the clearance to the back post while keeping the overall size of the machine small. The additional clearance is useful when swinging stock to cut curves. The trade-off is that three-wheelers can't handle the wide blades and higher tensions needed for resawing thick stock. These machines also can be harder to align properly and are more likely to have the blade crack from metal fatigue, a problem with all machines that have small wheels.

MOVING THE BLADE

On all saws, the lower wheel is driven by the motor. Some machines use a pulley and belt-drive system, while others use a direct drive, where the lower wheel is mounted directly on the motor's shaft. The axle of the upper wheel is mounted on a casting that slides vertically in the frame to tension the blade. The tension is controlled by a heavy spring compressed by a nut on a threaded rod, and the degree of tension is typically shown on a calibrated scale (see the top photo on p. 140). A second adjustment allows the upper wheel to be tilted several degrees to either side of vertical. This is called tracking, which along with a crowned tire keeps the blade centered on the face of the tire (see the illustration on p. 140).

GUIDING THE BLADE

To stabilize the blade, guide assemblies are mounted above and below the table (see the bottom photo on p. 140). The upper guide is on an adjustable post so the guide can be positioned just above the surface of the board. Guides support the blade on both sides and on the back edge. The back support is called the thrust bearing, which is typically a ball bearing, although some European machines use a steel disk running in a sintered-bronze bearing. Depending on its design, the blade can run against either the side or the face of the thrust bearing. On better-quality guides, the bearing is covered by a hardened steel face to resist wear.

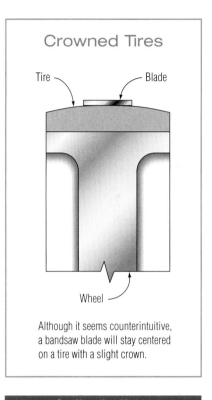

Crowned Tires

Tire — — Blade

Wheel

Although it seems counterintuitive, a bandsaw blade will stay centered on a tire with a slight crown.

Cooling the Motor

Until recently, most consumer-model bandsaws were underpowered, equipped with motors of only ½ hp or ¾ hp. Newer saws have 1-hp motors, but this may be inadequate if you plan to resaw wide planks. If you do upgrade to a larger motor, buy a totally enclosed fan-cooled (TEFC) motor, which won't become plugged up with sawdust and overheat.

Blade tension is controlled by the handwheel at the top of the vertical rod. The knob below it is used to adjust blade tracking.

Solid blocks in this guide assembly support the sides of the blade; a bearing supports the back as pressure is applied during a cut.

Blade Tracking

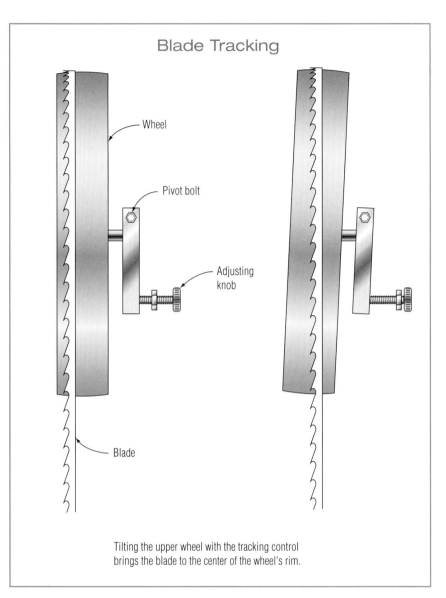

Tilting the upper wheel with the tracking control brings the blade to the center of the wheel's rim.

The two side supports can be either ball bearings or blocks of various materials that simply pinch the blade between them. Ball bearings need less daily maintenance than blocks, but they can still wear out. They also are fussier to adjust and may need frequent cleaning when cutting woods that contain a lot of pitch.

Block-type guides (see the sidebar on the facing page) offer simplicity, easy maintenance, and excellent blade support. They are used on both consumer- and industrial-grade machines. Several suppliers sell aftermarket ball-bearing guide assemblies to replace block guides, but they generally don't improve a saw's performance and may, in fact, degrade it if not carefully adjusted.

Blade Guide Blocks

GUIDE BLOCKS COME IN A DOZEN DIFFERENT MATERIALS, and each supplier claims its own brand is the best. In reality, almost all blocks will work well if their faces are kept flat and square and they are adjusted to a close fit on the blade. Blocks fall into two main categories: hard types, which are made from steel or ceramic, and soft types, which are made from brass, wood, or plastic.

Hard blocks have one main advantage: Their faces stay flat because they wear slowly. This allows you to set them once and then do a lot of sawing without having to readjust or reface them. They also have two disadvantages. First, they are more difficult to reface once they have become grooved. Second, they will ruin your blade instantly if they strike the teeth. Hard blocks are best used on wide blades where they can stay well clear of the blade's cutting edge.

Some suppliers claim that hard blocks create so much friction that they overheat and damage the blade, but I doubt it. Far more heat is generated by the friction of sawing. Steel blocks come as original equipment on many saws and, as long as you are careful, they give excellent service. I set hard blocks to bear very lightly against the side of the blade and then rotate the full blade through them by hand to make sure that a bad weld or kink won't cause jamming.

Soft blocks come in a wide variety of materials. They will not damage the blade's teeth, and they support the blade well because they can be fitted tightly—right around the teeth on very narrow blades. The softest of them are made of pure plastic, probably a nylon, and wear far too quickly to be practical. I've also read reports suggesting these soft blocks can melt and gum up the blade. A composite block called Cool Blocks® work better and won't melt, but they still wear rather quickly.

Brass or bronze both make excellent block material and strike a good balance between wear resistance and ease of refacing. They are easy to file flat when they develop grooves. These are used on larger machines, but I am not aware of anyone supplying soft metal blocks for smaller Delta-style saws.

Wood blocks are my favorite. They wear well, are easily replaced for free from the scrap box, and can be made up by the dozen quickly. This last point is a big advantage. Any time the block is worn in the slightest, you can just throw in a fresh set and get back to work quickly.

For angled cuts, a bandsaw's table tilts on curved supports called trunnions. On many saws, the table tilts to both sides of vertical, which is useful when cutting dovetails.

Troubleshooting

Bandsaws are susceptible to three problems: poor-quality cuts, vibration, and poor blade tracking. Most cutting problems can be solved by choosing the correct blade and fine-tuning the saw; vibration and tracking problems usually require repair or replacement of parts.

Understanding Blade Bow

A bandsaw only cuts well when its blade is perfectly straight, and almost all cutting problems with the saw are a result of the blade bowing under the stress of sawing. To understand the dynamics of bowing, try a small experiment. Cut a ¾-in. by 6-in. strip of paper and pull it taut between your hands. Reaching with a finger, apply pressure to the edge of the strip at its midpoint and notice what happens. The strip refuses to bow straight back; it flexes sideways in a broad curve.

A barrel-shaped saw kerf is the trademark of a blade that bows under pressure during a cut. A sharp blade of the right type and the correct tension will correct the problem.

Choosing a Blade

THERE'S A LOT TO CONSIDER WHEN CHOOSING A BLADE, most important of which is blade width. If you aren't going to be cutting a curve, choose the widest blade your saw can tension properly. That may be only a ½-in. blade on home-shop saws. Industrial machines may be able to handle blades of up to a couple of inches. For curved patterns, you'll need a narrower blade. A ⅜-in. blade typically can turn a curve with a radius of 1½ in., but any cut will go easier if you don't push a blade to its limit.

Next in importance is the number of teeth per inch (tpi), which is also called pitch. A good rule of thumb is to keep from 6 to 12 teeth in contact with the stock while you cut. Choose a coarser blade for fast cutting and a fine blade for a smoother cut. Softwoods are best cut with a coarser blade because it clears the sawdust better.

Some top-of-the-line blades have a variable pitch, meaning the teeth are alternately closer and farther apart to prevent the blade from developing a steady vibration that produces a rougher cut. A typical tooth spacing on a variable-pitch blade is marked as 2/3 pitch, which means 2 to 3 teeth per inch. A skip-tooth blade achieves a wider tooth spacing with small teeth by leaving out every other tooth.

The face of a saw's teeth can either be cut square to the blade, a 0-degree rake, or hooked with a positive rake of 10 degrees or more. Hooked teeth cut faster but rougher than standard or skip-tooth blades, which are sharpened with a 0-degree rake.

The last decision about a blade is the metal it is made from. This is as much a pocketbook decision as it is a technical one. Carbon-steel blades are the least expensive. They cut well but dull quickly if they get hot. These blades are best for light cuts in softer woods; don't expect to rip rosewood with them.

Bimetal blades have teeth of high-speed steel, whereas the main part of the blade is ordinary steel. High-speed steel won't dull like carbon steel when it gets hot, and the teeth will stay sharp much longer when cutting dense hardwoods. Bimetal blades cost considerably more than carbon-steel blades, but their longer life span—makers claim 10 times longer—make them more economical over time.

Carbide-tipped blades for smaller bandsaws are just becoming popular. At more than $100 a blade, they would appear to be a luxury, but for resawing wide hardwood boards, they do cut superbly fast, smooth, and straight. If you intend to do a lot of resawing in difficult woods, they are worth the investment.

Under the pressure of cutting, a bow can form in a bandsaw blade. Several things, all of them bad, result. First, the sides of the kerf take on the sideways bow of the blade, the infamous barrel-shaped cut that appears when sawing thick stock (see the photo at left). Second, the blade repeatedly wanders to one side and then pulls back in. This makes it very difficult to follow the intended line of the cut, and it creates a washboard surface on the face of the board. Third, the bowed blade binds in the cut, increasing the load on the motor and overheating the blade.

A couple of things can help. For straight cuts, always use the widest blade your saw can tension properly. For sawing curves, use the widest blade that will negotiate the tightest curve in your pattern. The type of teeth on the blade, their spacing, and sharpness also are critical (see the sidebar on the facing page). The wrong blade or a dull blade will need higher feeding pressure, which makes the bowing problem worse.

BLADE TENSION

A critical element that affects blade stability is tension. For demanding work, such as resawing, blade makers recommend that a blade be tensioned to 30,000 lb. per square inch (psi) of its cross section, which translates to a pull of 375 lb. on a ½-in.-wide blade 0.025 in. thick. The saw's frame, bearings, and wheel will actually have to cope with twice that load because both the cutting and returning part of the blade are equally taut. Most home-shop-grade tools simply can't sustain this much strain without damage, and their springs and tensioning scales will often limit you to less than a quarter of the recommended tension. Manufacturers generally recommend blades on these machines be tensioned to a maximum of 15,000 psi. A complicating factor is that the tension spring weakens with use, making it even harder to properly tension the blade. Replacing the spring every few years is the only solution.

There is no simple way to measure blade tension, since tension gauges built into the bandsaw aren't very accurate. The best way to gauge the tension of a blade is to measure how much the blade stretches as it is tightened. The amount is small, but a 5-in. length of blade will stretch 0.001 in. for every 6,000 psi of tension that is applied. Starrett® and Lenox® both sell gauges that measure how much the blade stretches when it's being tightened, but they cost several hundred dollars (see the photo at right).

The easiest approach for home-shop bandsaws is to tension the blade to a higher setting on the saw's built-in scale. In other words, tighten a ½-in. blade to the ¾-in. setting. It is not a good idea, however, to exceed the highest tension setting on the saw's scale. Too much tension may crush the spring, which limits its role as a shock absorber, and can cause wheel and bearing damage.

Another option is to make a simple device that measures blade tension (see the sidebar on pp. 144–145). You can make this tool in an hour from a scrap of hardwood and two nails. It attaches to the bandsaw blade with two clamps 5 in. apart. As the blade is tensioned, the gap between the upper and lower halves of the tool increases, so you measure the change by slipping a feeler gauge between the measuring pin and the lower block. A second pin goes into both blocks to keep the two parts of the assembly lined up. The tool works on all blade widths.

Using the tool is simple. Install a blade on your bandsaw, tension it lightly, then adjust the tracking and the guides while moving the blade by

Give the Springs a Rest

Tension springs weaken with use, but they last longer if you reduce the tension on the blade when the saw isn't being used. To prevent the blade from slipping off the wheels, don't completely remove the load from the spring. Just back it off to the lowest setting on the tension gauge.

Although very accurate, blade tension measuring devices are too expensive for most home shops.

A Simple Tool to Tension Bandsaw Blades

A FEW HARDWOOD SCRAPS, TWO NAILS, and a little time are you'll need to make a blade-tensioning device that is surprisingly accurate on any width and any type of blade. The idea is simple: Blades stretch slightly as they are tensioned. By measuring the gap between the two halves of the device as blade tension is increased, you can determine just how tight the blade is. This approach is much more accurate than using the tension scale built into the saw—and much cheaper than buying a tension measuring device that does essentially the same thing.

2 In one end of the block, drill a ³⁄₃₂-in. hole as deep as the bit will allow for the alignment pin. Clamping the block upright on a shopmade V-block makes it easy to drill the hole accurately.

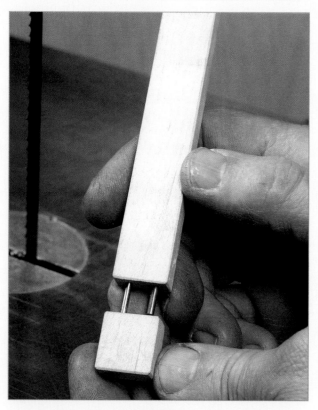

1 Made from a scrap of hardwood and two finish nails, this simple gauge allows accurate tensioning of a bandsaw blade by measuring how much the blade stretches. Start with a scrap of fine-grained hardwood ⅜ in. by ¾ in. by 5⅜ in. long.

3 Before making a cut, make index marks on both sides of the cut line so you can line up the two pieces in their original orientation. Carefully cut a ¾-in.-long piece from the drilled end of the block, then polish the cut face of the smaller block by running it across very fine sandpaper on a flat surface.

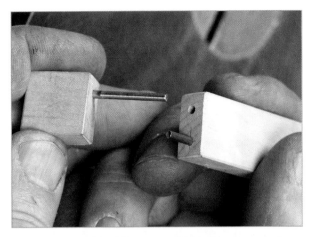

4 With a ⁹⁄₃₂-in. bit, drill a second hole ½ in. deep in the end of the long block next to the alignment pin hole. Make the measuring pin by cutting a 6d finish nail ⅞ in. long. To prevent snagging the feeler gauge, slightly round the working end of the pin and tap the pin into the hole you drilled for it. An alignment pin 1⅝ in. long, also cut from a 6d nail, fits in the other hole.

5 Before inserting the alignment pin, file one end slightly to reduce its diameter and allow it to slide easily. Chucking the pin in a drill press is an easy way to file it, but don't take off too much material.

Measuring Blade Tension

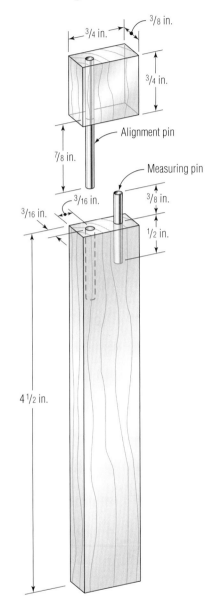

3/4 in.

3/8 in.

3/4 in.

Alignment pin

7/8 in.

Measuring pin

3/16 in.

3/16 in.

3/8 in.

1/2 in.

4 1/2 in.

This shopmade device consists of two finish nails and a small block of hardwood, but it accurately measures blade tension on a bandsaw.

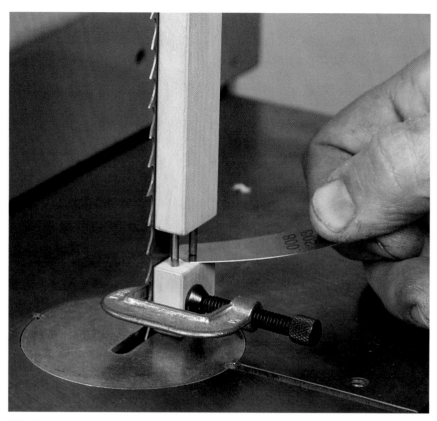

With the gauge clamped to the blade and tension increased, a feeler gauge measures the gap under the pin. Each 0.001-in. increase in the gap indicates an additional 6,000 psi of tension.

hand. Once the blade rotates properly, power up the saw and let it run for a minute or two to warm up the blade and tires, then cut the power and unplug the saw. Next, back off the tension until there is just enough pull on the blade to prevent it from going slack and slipping off the tires; normally this will be a little below the tension mark for a ⅛-in. blade on the machine's scale.

Raise the upper blade guide out of the way, and lightly clamp the jig to the side of the blade just behind the teeth. The measuring pin should be toward the back of the blade. Loosen the upper clamp and pinch a 0.005-in. feeler blade between the measuring pin and the lower block. Leaving the gauge in place, tighten both clamps.

Next, pull out the feeler gauge and recheck the gap. It may change slightly from the torque of the clamps, but it isn't important that it be exactly 0.005 in. Using the gauge set, find and note which gauge just slides between the pin and the block.

To tension the blade, begin by choosing a feeler gauge the width of the starting gap plus an additional 0.001 in. for each 6,000 psi of tension you want to place on the blade. For 15,000 psi of tension, add 0.003 in.

to the initial gap, which will get you to 18,000 psi. Once there, you can back off the saw's tension adjustment a little to get in the neighborhood of 15,000 psi.

With the correct feeler gauge in hand, increase the blade tension while checking the gap under the pin. When you reach the correct tension, the gauge will slide under the pin. Mark your saw's tensioning scale at its pointer, noting the width of the blade. The recalibrated scale will allow you to quickly tension the same width blade in the future without using the tool every time.

If you build this tool and try it on any small, consumer-grade machine, such as the Delta 14-in. saw or its numerous clones, you will probably discover that the saw can't even come close to the recommended 15,000 psi tension. With a ½-in. blade in the saw, you are likely to find that running the indicator up to the ¾-in. blade setting increases the gap by only 0.001 in. (6,000 psi) or not at all.

ADJUSTING GUIDES AND BLOCKS

Shortening the length of the blade between the guides to a minimum also will reduce bowing. Always set the upper guides so they are close to the surface of the stock. You can't adjust the lower guides, but the best-designed machines have them tucked up close to the underside of the table. Block guides often work better because their compact design allows them to move in closer to the stock than the bulkier ball-bearing guide assemblies.

Of course, the most basic adjustment may be the most important to prevent bowing: Make sure the guide blocks fit tightly to both the sides and the back of the blade. And, here again, simple block-type guides often work better than ball bearings, which need to be set a bit looser to cut down on noise and wear. You should check blocks regularly for wear and resurface them as needed. In this regard, wood blocks are easy to service because the tools and materials for fixing and replacing them are always at hand.

Wood guide blocks are simple to make and work very well. The ideal material for the guides is lignum vitae, a dense tropical wood with natural oils that was used extensively as a bearing material in old machinery. A good substitute is hard maple with an added lubricant, such as DriCote® spray-on blade and bit lubricant made by Sandaro Industries. It is commonly available from many woodworker-supply companies and it works very well in this application. Simply cut the wood to the right size, then spray the bearing end of the block before you install it in the guide holder.

Other than poor cutting, the most common problem with a bandsaw is vibration. In addition to the usual suspects—bearings, belts, and pulleys—wheels and especially the tires are often at fault. A wheel can be warped because the casting was badly done or because it was bent in shipping or by severe overtensioning of the blade. A second source of vibration is poorly balanced wheels, which is fairly common on low-budget machines.

Replacing the Spring

If your saw can't reach 15,000 psi of tension, it's because the spring on the machine has been crushed. It will exert far less pressure than it was originally designed for, no matter how far it is compressed. All you can do is buy a new spring, preferably one of the better-quality aftermarket ones made by Iturra Design of Jacksonville, Florida, which you can call at (888) 722-7078. Made from a better grade of steel, and more of it, the Iturra spring won't weaken over time.

On occasion, a well-balanced wheel can be thrown off by clumps of gummy sawdust packed onto the inner surface of the rim. A buildup of sawdust and pitch on the surface of the tire also can lead to vibration. And, as a machine gets older, the tires can harden, crack, and eventually break away, leading to moderate to severe vibration.

A final problem with bandsaws is the failure of the blade to track correctly. It may wander or walk entirely off the rim while cutting. This is usually due to grooves in the tire or wear that reduces the crown profile. Tracking problems also can be caused by looseness or binding in the upper axle and tracking assemblies or by large misalignments between the upper and lower wheels.

Set Up, Then Tune Up

Tuning up a bandsaw is very different from tuning a table saw. On a table saw, you should check adjustments for square and parallel a few times a year; in between, very little needs to be done. On a bandsaw, you only have to do a number of adjustments and minor repairs once, in a process called setting up the machine. Afterward, you'll make small adjustments on an almost daily basis, similar to a musician tuning up an instrument each time he plays. The main purpose of the setup is to make the various controls for adjusting the saw work so smoothly and reliably that daily tune-ups are effortless.

The best approach is to tackle one part of the bandsaw at a time. Although the machine may appear complicated, its individual components are fairly simple. Take your time, keep track of the parts, and take notes or make a sketch to help with reassembly.

The setup procedure tackles parts of the machine in the following order:
- Bearings, belts, and pulleys
- Wheels and tires
- Upper wheel tension and tracking assembly
- Frame joint
- Table trunnions
- Lower blade guides
- Wheel alignment
- Guidepost and table alignment

At each step, you'll be cleaning and smoothing parts, removing paint, applying grease and oil, and adjusting alignments. Much of this may seem tedious and trivial, but the end result will pay off.

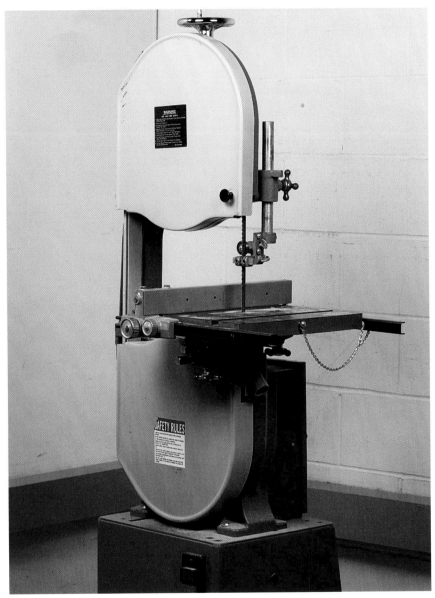

Widely imitated, the Delta 14-in. bandsaw has been in production since the 1930s with only minor changes.

The bandsaw I've used to explain the process is a Delta 14-in. model, a tool that has been in continuous production for more than 60 years with only minor design changes. This saw, and its Taiwanese clones, is by far the most common bandsaw in woodworking shops. The Delta is well designed, but it can be greatly improved by going through the following setup and tuning procedures. If you have a European or industrial-grade saw, some of the steps may not be needed, but any bandsaw will benefit from cleaning, lubricating, and aligning.

On some saws, tire replacement is easy: The tire just snaps into a groove in the wheel rim. Replacements on other saws, however, must be glued on.

Sanding the outer edges of a tire can restore the crown and correct blade-tracking problems. Tires on Delta-style saws are thinner and should not be sanded.

Setting Up a Bandsaw

Begin by unplugging the saw and removing the blade. Check the motor, belts, pulleys, and bearings for problems. Sight and hearing will give you many cues. A misaligned pulley makes a lot of noise, and you can actually see the excess vibration in the pulley as the saw runs.

WHEELS AND TIRES

Carefully examine the condition of the tires on each wheel. Using an abrasive pad, remove any sawdust and gum. The combination of heat and pressure from the sawblade can make the sawdust adhere tenaciously, but don't use solvents: There is a chance they will damage the rubber.

The tires should be smooth and flexible, without any cracks or grooves in the rubber. If the tires have deteriorated, you'll have to replace them, which is easy on Delta-style saws. The tire sits in a groove machined in the wheel's rim, so the tire just snaps into place (see the photo at left above). On most other saws, you'll have to glue the new tire onto the rim with rubber cement or auto weather-strip adhesive.

Most tires should have a crown of about $\frac{1}{32}$ in., although some large saws designed to use wide blades have flat tires. If the crown is worn away but the tires are otherwise in good shape, you can restore the crown by sanding it. To restore the crown on the lower tire, run the wheel under power, with the blade removed, while holding a sanding block against it. Pivot the block to create the right shape (see the photo at right on the facing page). To restore the upper tire, have a helper spin the wheel either by hand or by holding a sanding drum chucked in an electric drill against it while you work with the sanding block. Sanding rubber will create an obnoxious black dust, so try to keep it out of the air by picking it up with a shop-vacuum hose held next to the block while you sand.

On Delta-style machines, however, don't try to restore the crown by sanding; the tire is too thin. There is a crown profile machined into the wheel's groove, so replacing the worn tire solves the problem.

TENSION AND TRACKING CONTROLS

To service the tensioning and tracking controls for the upper wheel, the wheel must come off (see the photo at left below). Slide it off its axle after unthreading the nut on the end of the shaft (keep track of all the washers and where they go). Remove the wheel cover, then slide the entire axle assembly out of the grooves in the saw's frame. Clean off the old grease and pass a file over the sliding surfaces to remove any rough spots, then examine the small casting that the axle is attached to, making sure it isn't cracked or bent. On Delta-style saws, the axle may be a very loose fit in the casting but this is normal; the tension on the blade will tighten it up.

To check and lubricate the tension and tracking controls, you must first remove the saw's upper wheel. A nut at the end of the axle holds it in place.

The tension and tracking assembly is made up of only a few simple parts, but they must work very smoothly if the saw is to perform properly.

The tensioning and tracking control handles on some bandsaws are small and awkwardly located. After-market suppliers offer replacement handles that are better designed and much easier to use.

Removing the paint from the machined surfaces at the joints of a saw's frame will dampen vibration and make it easier to align upper and lower wheels accurately.

If you find any paint in the grooves of the saw's frame, which is not uncommon, you should scrape it out using the tip of a utility knife, then finish up with a pass or two of a file. Paint in the grooves will make the tension control bind and cause the saw to perform erratically.

If the saw is more than two or three years old or if it has seen a lot of use, this is a good time to replace the tension spring. The spring weakens dramatically over time, making it impossible to properly tension the blade.

The last thing to do before reassembly is to round off the end of the threaded tension rod and the smaller track-control rod so that they turn smoothly. Apply oil to the hinge pin of the tracking mechanism and use grease or stick lube on all of the sliding and threaded parts before sliding the axle assembly back in the saw frame. The ball bearings in the wheel hub are typically sealed and can't be lubricated.

Before reinstalling the upper axle assembly on Delta-style saws with two-piece frames, consider unbolting the upper frame casting from the base casting and scraping the paint off their mating surfaces (see the photo above). Paint between the two parts throws off the alignment of the band-saw, and it also "softens" the joint, making the saw more likely to flex and vibrate under load. This problem is even worse if the machine has a riser block because there will be a corresponding increase in the paint layers between all the parts.

UPPER BLADE GUIDES

Remove the guide blocks or bearings and check their condition. The bearing surfaces of blocks should be smooth and square, so grind, sand, or replace them as needed (see the photo at left below). While the blocks are out, check that the metal casting in which the blocks are mounted hasn't been distorted or cracked from overzealous tightening of the set screws. It may need to be replaced.

Any ball bearings used in the guide should turn smoothly, without any grittiness. You can try to free tight bearings using a dose of light machine oil, but any success may be short lived and the bearing probably will have to be replaced. If the front face of the thrust bearing is badly chewed up, replace the bearing. Clean off any pitch and sawdust on the rim of the bearings using penetrating oil and abrasive pads.

Next, clean up the threads of all the thumbscrews or lock bolts in the guide assembly and slightly round their tips so they'll hold better and adjustments won't creep as they're tightened. Use lubricant on all of the moving and threaded parts of the guide as you reassemble it.

On Delta-style saws, if you look at the end of the hex-shaped rod that supports the thrust bearing, you'll see that the bolt hole is off center. Install the rod in its socket so that the center of the bearing is as far as possible from the blade. The blade should touch only near the rim of the bearing (see the photo at right below).

Refacing worn guide blocks is easy if they're made from a soft material such as wood. Faces should be flat and square to the sides.

The upper thrust bearing on a Delta-style saw is mounted off center on the end of a hex-shaped rod. Install the rod so the center of the bearing is as far away from the blade as possible.

Paint on the mating surfaces of the table trunnions may cause binding. Rarely serviced, the trunnions work much better after they are cleaned and lubricated.

Clean up the guidepost and remove any burrs on it using a file. If needed, shape the end of the post-locking screw to fit the groove of the post so it will hold better. Later, when a blade is installed, the position of the guide assembly on the end of the rod will be adjusted to center it on the blade.

TABLE TRUNNIONS

On smaller saws, the next step is to remove the table by unthreading the handles on the bolts that lock the trunnions. The table should then lift right off.

From the underside of the table, unbolt the upper halves of the trunnions. This will allow you to check how the heads of the lock bolts fit inside their grooves in the trunnion castings. Often they will fit poorly and the soft metal on the inside of the castings will be chewed up. Spend a few minutes using a file to clean up the damage and get the bolt heads to fit properly. Use a stick lube on the bolt heads and all of the sliding surfaces when you reassemble the trunnions.

Bolt the trunnion upper halves back on the table but leave the bolts slightly loose. You should tighten them up only after you replace the table on the saw's base and the upper and lower halves of the trunnions are mated.

On larger saws, the table can weigh several hundred pounds, so it makes a lot more sense to work on the trunnions with the table in place. If the curved surfaces of the trunnions are painted, carefully scrape the paint off and polish the bearing surfaces with emery paper.

Bathed in sawdust and resin, the lower blade guide needs occasional cleaning and lubrication to keep it working smoothly.

LOWER BLADE GUIDES

Before replacing the table, you should work on the lower blade guide assembly while it is easy to get at. As with the upper guide, check and repair or replace the blocks and bearings as necessary. If the table is in place, tilting it will often improve your access to some parts of the lower guides.

The mechanism that adjusts the lower guides on Delta-style saws is a bit complicated, so take it apart for a thorough cleaning only if it binds after cleaning and greasing it in place. Remove the old grease and sawdust using a toothbrush and penetrating oil while moving the parts back and forth with the adjustment knobs, then grease the parts with a stick lubricant. If you wind up taking the guide mechanism apart, lightly file all of the sliding surfaces and don't lose the small springs between the wedges that clamp the guide assembly to its mounting plate.

WHEEL ALIGNMENT

When aligned correctly, the wheels are in the same plane (see the illustration at right). To check, place a straightedge across the rims of both wheels at the same time. On Delta-style saws, this can be easily accomplished with the table in place, but on many other saws the table must be removed, a daunting job on a big machine with a heavy, cast-iron table. A second problem with some machines is that the wheels arc set back in the frame and the straightedge can't reach the rims unless it's notched to clear the wheel housing. A shopmade plywood straightedge is helpful.

Lay the straightedge across the rims of both wheels, adjusting the tilt of the upper wheel to bring it parallel to the straightedge (see the photo at left on p. 156). Ideally, both edges of both rims will touch the straightedge.

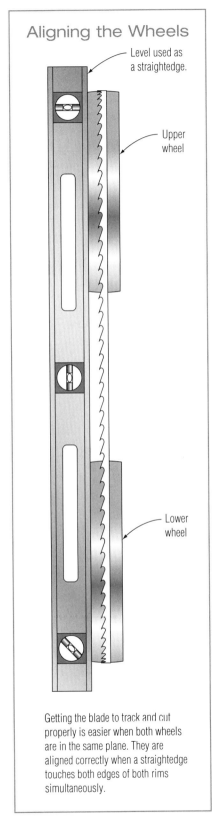

Aligning the Wheels

Level used as a straightedge.

Upper wheel

Lower wheel

Getting the blade to track and cut properly is easier when both wheels are in the same plane. They are aligned correctly when a straightedge touches both edges of both rims simultaneously.

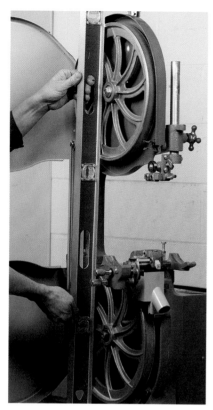

A straightedge tests whether the upper and lower wheels run in the same plane, which is crucial for reducing wear on the saw's tires and reducing stress on the blade.

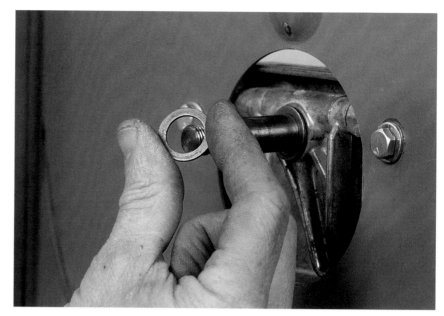

On most saws, bringing wheels into the same plane simply means adding or subtracting washers on the axle between the wheel hub and the frame.

If they do not all touch, one of the wheels will need to be adjusted, in or out, by adding or subtracting shim washers on its axle as shown in the photo above right). On most Delta-style saws, the upper wheel is adjusted, whereas on Jet and some Craftsman saws it is the lower. One aftermarket supplier offers a kit of graduated shim washers for Delta bandsaws and its clones to make adjustments a simple task.

Once the wheels are aligned, install a medium-width blade (a ⅜-in. blade is fine) on the saw and lightly tension it. With the power still disconnected, turn the upper wheel by hand as you fully tension the blade and adjust the tracking to center the blade on the upper wheel, making sure that the guides are fully backed off and not touching the blade.

With the blade centered on the upper wheel, check that it's centered on the lower wheel also. If it isn't, you may want to readjust the wheel-alignment washers to compensate for the flex of the saw's frame and wheels under load.

LINING UP THE GUIDES AND POST

If a bandsaw is to cut properly, the blade guides must be set correctly, and the setting should stay the same as the guide assembly is moved up or down to accommodate stock of different thicknesses. A misaligned guide-post throws the settings off as the assembly is moved, so an important part of a tune-up is to make sure the post is straight.

On saws where blade guides can be adjusted, shift the entire upper and lower blade guide assemblies to center them on the blade's path (see the photo at left below). Next, lower the guidepost until it is close to the table, then move one of the side bearings or blocks in until it almost touches the blade (all the other blocks and bearings should still be pulled back). Now raise the post to its highest setting and check that the block still just clears the blade, which would mean the post is properly aligned and moving parallel to the blade. If the gap either disappears or grows larger, it means the post is out of line. Unless it is realigned, you will be forced to readjust the guides each time you change the height of the upper blade guide—an unnecessary pain in the neck.

On some saws, the guidepost mounting bracket can simply be loosened and shifted to get the post properly aligned. On Delta-style saws, however, the only way you can adjust the post alignment is by tilting the entire upper frame casting. This is done by inserting a shim between the upper frame and the base casting, which isn't that hard to do. Cut the shim about ⅜ in. wide and a bit longer than needed, then install the shim by loosening the bolt holding the frame together and slipping the shim into the joint (see the photo at right below). Cut off the excess shim after retightening the bolt. You'll need to determine the proper shim thickness by trial and error. The last time I did this, a piece of aluminum flashing 0.010 in. thick worked perfectly. If you changed the post alignment, check that the upper guide assembly is still centered on the blade path; you may need to readjust it.

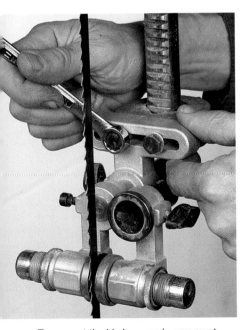

To support the blade properly, you must adjust the entire guide assembly from side to side so it is centered on the blade's track.

Inserting a shim in the joint between upper and lower frame pieces may be the only way to keep the guidepost straight as it is adjusted up or down. The procedure is easy.

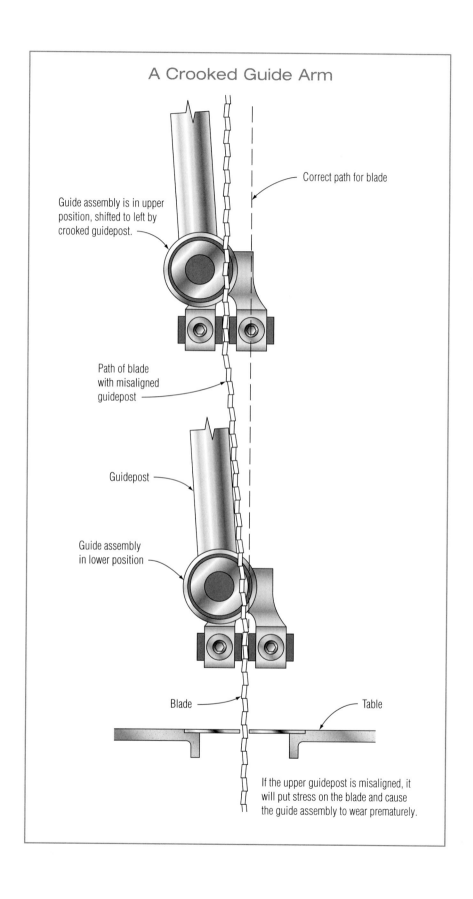

A Crooked Guide Arm

Correct path for blade

Guide assembly is in upper position, shifted to left by crooked guidepost.

Path of blade with misaligned guidepost

Guidepost

Guide assembly in lower position

Blade

Table

If the upper guidepost is misaligned, it will put stress on the blade and cause the guide assembly to wear prematurely.

SQUARING UP THE TABLE

The only remaining setup procedure is to square the table to the blade. Begin by backing off the blade guides and adjusting the blade to full tension and proper tracking in the center of the wheels. Place a square on the side of the blade and rotate the table in the trunnions until the blade is square to the table, then set the 90-degree stop if your table has one. Once the table is square, adjust the pointer or move the tilt scale to read zero (see the photo below). Later, when you make test cuts, you can fine-tune this setting.

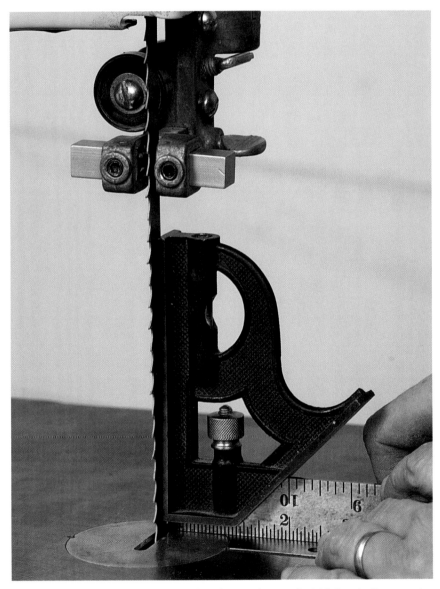

One of the last steps in setting up a bandsaw is to make sure the table is actually square to the blade when the pointer reads 0 degrees on the tilt scale.

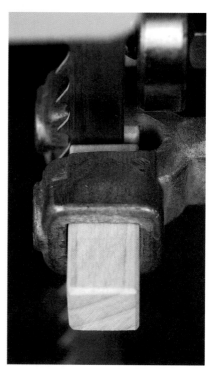

On wider blades, the guide block or side bearing is positioned to touch the blade at the base of the gullets.

The last alignment to check is whether the back of the blade is square to the table, although this is important only if you are going to be cutting joints such as tenons or dovetails on the saw. If the blade isn't square, the only fix on most saws will be to insert shims between the trunnion-support casting and the base of the saw.

That's it. Setting up a bandsaw is a long process. But now the saw should need only routine maintenance, which mostly amounts to cleaning and lubrication, although the guide blocks and bearings will occasionally need repair or replacement. The next section will explain how to adjust the guides to the blade you are using and line up the fence and miter gauge.

Tuning Up a Bandsaw

Once the bandsaw is properly set up, the adjustments that are needed for a tune-up will go quickly. Begin by unplugging the machine and installing a good blade for crosscutting the 1-in. stock of the test board. A 10-tpi or 12-tpi, ⅜-in.-wide blade is a good choice. Before installing the blade, pull back the guide blocks and bearings far enough so they won't touch the blade while you adjust the tension and tracking.

INSTALLING A BLADE

While turning the upper wheel by hand, alternately adjust the tension and tracking controls to get the blade to full tension and centered on the upper wheel face (if you've just finished setting up the saw, you will already have the blade in the machine). On small machines, where the tension gauges typically read low, tension the blade to the next highest setting. I then usually turn on the machine and let it run for a minute to make sure the blade is settled in and tracking properly. Once the blade is set, it is time to position the guides.

ADJUSTING THE GUIDES

Always unplug the power before adjusting the guides because your fingers are just too close to the blade if the machine should accidentally start. Begin by lowering the upper guide until it will clear the stock to be cut by ¼ in. to ½ in. At this point, none of the blocks or bearings, above or below the table, should be touching the blade. From here on, you can fully adjust one guide assembly at a time or move back and forth between them for each step.

Next, move the entire guide assembly forward or back until the leading edge of the side blocks or side bearings will touch the blade just behind the deepest part of the gullet (see the photo at left). If needed, bring one block close to the blade to check the position. When the alignment is right, lock its set screw.

Bring the thrust bearing forward until it either just touches the back of the blade or there is a very small gap between it and the back of the blade as shown in the photo at right). When it's set right, lock its set screw.

Now bring one side block or side bearing in to just touch the blade. Keep pushing the blade toward the block with a series of light touches. This will allow you to feel the gap closing, which is often easier and more accurate than setting the gap by eye. When the last bit of flex disappears, the block or bearing is in the correct position and you can tighten it down (see the photo below).

The final step is to position the second side bearing. If the block is soft—made of wood, brass, or plastic—you can lightly pinch the blade against the opposing block and lock it in place (see the top left photo on p. 162). If the block is made of steel or ceramic or if it is a ball bearing, leave a slight gap. Pinching a piece of paper between the blade and the block or bearing before tightening the set screw will give you the right setting (see the top right photo on p. 162).

Once both the upper and lower guides are set, turn the upper wheel by hand through several revolutions to make sure there are no kinks or a poorly ground weld on the blade that will bind as it passes through the guides. If moving the blade by hand goes well, restore power and turn on the machine for a final check.

With the machine running, you may find that the ball-bearing guides are spinning. You might want to back them off a bit more if you don't want them to turn when you aren't feeding stock, but this increases the chance that the blade will bow and cut less accurately. If you back off the

Move the thrust bearing so it lightly touches, or nearly touches, the back edge of the blade. During a cut, the bearing prevents the blade from moving backward.

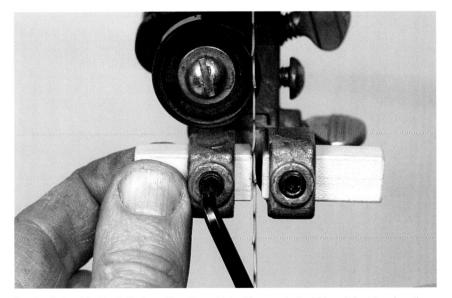

Set the first guide block (it doesn't matter which side you start with) so it just touches the blade but does not push it to the side.

Blocks made of ceramic or steel, as well as ball-bearing guides, should be adjusted so there is a very slight clearance between block and blade. A piece of paper sets the distance and is removed once the block's set screw is tightened. If blocks are made from a soft material, you can press the second one lightly against the side of the blade and then lock it in place.

If blocks are made from a soft material, you can press the second one lightly against the side of the blade and then lock it in place.

Narrow blades may be fully enclosed by soft guide blocks to increase support and reduce bowing or twisting. Hard blocks would ruin the blade when used this way.

thrust bearing, be careful that the blade can't push back so far that its teeth are damaged by striking either hard blocks or the rims of the ball bearings.

On narrower blades, the procedure for setting the guides is the same until the blade width drops to less than ¼ in. On these narrower blades, you will get better blade support if you use soft-sided blocks that won't damage the blade's teeth and you position them to bear on the entire side of the blade including the gullets. On very narrow blades, you can completely enclose the blade. Bring the second block in as you move the blade by hand with the upper wheel so that the saw's teeth will cut their own clearance groove into the block faces (see the photo at left). The procedure for setting upper and lower guides is the same.

ADJUSTING THE MITER GAUGE

Once the guides are adjusted, you can set the miter gauge. Using a trued-up piece of 1x4, take a test cut with the gauge and check the cut face, both horizontally and vertically, with an accurate square. Adjust the stop on the miter gauge and the table trunnions to get 90-degree corners in both directions.

ADJUSTING THE RIP FENCE

The final adjustment is to line up the rip fence, which is done by making test cuts with the blade you will be using for ripping (you will have to readjust the rip fence each time you switch to a different blade). On Delta-style saws, use a ½-in.-wide blade, the widest that can be properly ten-

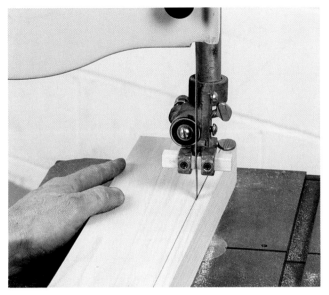

To adjust the rip fence, start by ripping a milled test board free-hand. When the saw is cutting to the line without course corrections, stop the saw.

Once you've found the lead angle, mark it on the table with a pencil.

sioned, and tension it to the ¾-in. blade setting. Choose a 4-tpi to 6-tpi blade for sawing a 2-in.-thick test board.

Start by setting up the saw with the blade fully tensioned and the guides properly adjusted. Draw a line down the length of the board, ½ in. from the edge, then rip the board freehand along this line with the wide side of the board on the side of the blade where you would normally mount the rip fence. As you follow the line, adjust the feed angle to find the orientation where the blade will naturally cut straight without drifting from side to side (see the photo above left). Once you have sawn a few inches without having to make corrections to stay on the line, hold the board in place and turn off the saw. Without allowing the board to shift, draw a pencil line on the saw's table along the back edge of the board (see the photo above right).

Remove the board, then mount the rip fence on the table close to the pencil line. Loosen the adjustment bolts for the fence bar and shift it to match the orientation of the pencil line, then angle it a few degrees more toward the blade before tightening the adjustment bolts (see the photo at right). The added toe-in creates a slight side pressure on the blade that dampens vibration and produces a smoother cut. As a final check, make a full-length rip on the test board. If the cut is straight and smooth, you are done with the tune-up. If the cut wanders or is very rough, check the blade tension and guide settings. If they're good, try changing the rip-fence alignment slightly.

The rip fence is lined up with the lead angle drawn on the table and then toed in toward the blade a few more degrees. The slight cant reduces blade vibration and makes for a smoother cut.

The Router Table

No other shop tool is as versatile as the router. It can cut complex joints and create an endless variety of moldings. Once it is mounted beneath a table, the router is transformed into a stationary tool of even greater capabilities. Like a table saw or a bandsaw, however, a router table will only produce the best work when its components are flat, straight, and square to each other.

Anatomy

Although a few metal router tables are available, they are generally small with fences that are difficult to use. They are a poor choice for furniture making. Much more practical are larger tables, typically 24 in. wide and 32 in. to 36 in. long, made with a core of MDF and faced on both sides with plastic laminate. A dozen or more companies make router tables with this basic design.

These tables all share a similar system for mounting the router. The tool's base is screwed to a panel that drops into a recess in the table's surface (see the photo at top right on the facing page). Most commercially made tables have adjustment screws to bring the panel flush with the table, as shown in the bottom photo on the facing page. Depending on the manufacturer, the panels may be transparent plastic, metal, or phenolic plastic. Transparent plastic panels will occasionally become distorted or sag under the weight of the router, whereas metal and phenolic panels rarely have this problem.

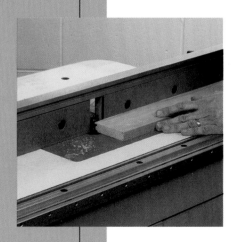

Nearly all full-sized router tables have settled into one standard design: a top made from MDF faced with a plastic laminate.

Mounted on a drop-in panel, the router is easy to remove from the table to change the bit.

Most tables have height-adjusting screws at each corner of the drop-in panel to bring it flush with the table's surface. This table has extra screws to prevent the panel from flexing.

Fences on commercially made tables also are similar. The backbone is an L-shaped aluminum extrusion with a face of wood, MDF, or plastic (see the photo at left on p. 166). On most fences the face is made in two sections that can be adjusted for a close fit around the router bit. Although this is an occasional advantage, fences with one continuous face are simpler to use and are less likely to catch the workpiece. Plastic-faced fences have a low-friction surface, but they often warp.

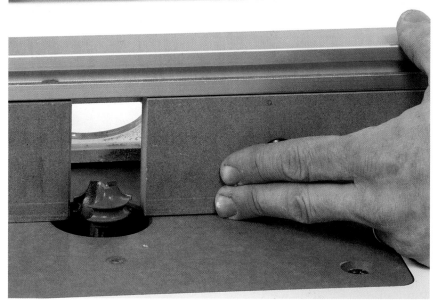

Many tables have a split fence that can be adjusted for a close fit around the router bit.

Almost all commercially made tables use an L-shaped aluminum extrusion for a fence; on the best-made fences, the metal is at least ¼ in. thick.

The Value of Inserts

Some drop-in panels have interchangeable inserts that can be switched to match the diameter of different bits. Most of these inserts, however, weaken support right where it is most critical. Moreover, inserts often are slightly higher or lower than the panel, and they may catch the workpiece as it passes. Unless you expect to use very large bits, a simple panel with no inserts and a 1-in. to 2-in. hole is a better design.

Unless the aluminum extrusions are fairly thick—¼ in. or more—the fence will be flimsy, and it can bow as it is being used. The best fences are machined by the manufacturer to an accurate 90 degrees; raw aluminum extrusions are never reliably square or straight.

Troubleshooting

For precision joint making and smooth molding runs, a router table needs to be reasonably flat and the fence must be straight and square to the table.

Flatness is surprisingly hard to achieve in any surface, but the typical router table has several strikes against it, starting with the material the table is made from. Although it appears rigid, MDF is really just high-tech cardboard. Because MDF has little strength, the center of a poorly supported table will inevitably sag. In addition to dishing in the top, most tables have a sharp kink at the miter-gauge slot because the laminate, which reinforces the top, is interrupted on the upper face of the table.

Unless carefully adjusted, drop-in panels and inserts often will be slightly higher or lower than the adjoining surface, causing stock to catch on the edge or drop into the recess. A few manufacturers intentionally crown their tables to ensure the stock is supported close to the bit. For most operations, a crowned table is better than one that is low in the center, but a table that is high in the center still creates problems because the workpiece rocks as it moves from the infeed to the outfeed side of the table.

Few commercially made tables stay flat in use, so the first step in tuning up a router table is to check the table for flatness. A builder's level is accurate enough.

Tuning Up a Router Table

Setting up and adjusting a router table is fairly simple and straightforward—no more than an afternoon's work. The only complex job is to flatten and reinforce the table, but not all tables will need this repair.

CHECKING THE TABLE FOR FLATNESS

A good carpenter's level is adequate for checking a router table for flatness; you won't need a precision straightedge. Begin by removing the drop-in panel, then check the table lengthwise and across its width, placing the level along the table's edges and over the center (see the photo above). Typically, you'll find a sag of ¹⁄₁₆ in. at the edge of the panel opening. If the table has a miter-gauge groove, the table will probably have a sharp kink along the length of the slot.

Use the level to check the table diagonally. If you turn up a bow along one diagonal but the other is flat or bowed in the opposite direction, the table is twisted—often the result of an uneven floor. You may be able to correct the problem by shimming the legs or readjusting the levelers.

FLATTENING A WARPED TABLE

Flattening a sagging table is a two-step process. First the sag has to be removed, then the top should be firmly attached to a reinforcing frame to prevent the sag from recurring. Begin by removing the table's top from its base and inverting it on a sturdy, flat benchtop or the top of your table saw. Support the corners on four blocks of even thickness. Next, place a rectangle of plywood a few inches larger than the panel cutout over the

Before reinforcing a warped tabletop with metal or wood, you should flatten it by flipping it over, blocking it up, and applying weight to the high side.

center opening and stack some weight on it—a few gallons of paint or its equivalent. In a few hours to a few days, the top will be flat again. Check it regularly.

If you overcompensate, flip the top over and push it back the other way. Flatten out any more localized kinks by shifting the blocks and weight to concentrate the force where it is needed. Once the top is flat, remove the weights and blocks and store the top on a flat surface until it is reinforced.

REINFORCING A TABLETOP

A wood or metal frame will strengthen the top and prevent it from warping again. One advantage to using angle iron for the frame is that a shallow frame is less likely to interfere with access to the router, and the frame will typically fit within an existing base cabinet. Another advantage is that a metal frame can be attached to the underside of the table with sheet-metal screws, leaving the top surface unmarred. There is no need to weld the frame; you simply screw lengths of angle iron to the underside of the table.

A wood frame made from ¾-in. plywood or MDF has the advantage of being inexpensive and easy to make. If braces are located carefully, a wood frame usually won't interfere with a router. Alternatively, you can use your existing base, if it is well made, to stabilize the top. Just add cross braces and attach the top with better hardware (see the photo at left).

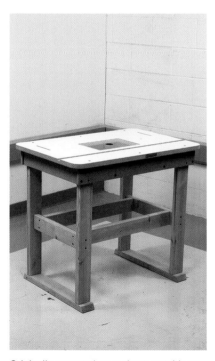

Originally mounted on a sheet-metal base, this rebuilt table is flatter and more stable now that it is attached to a wooden frame and leg assembly.

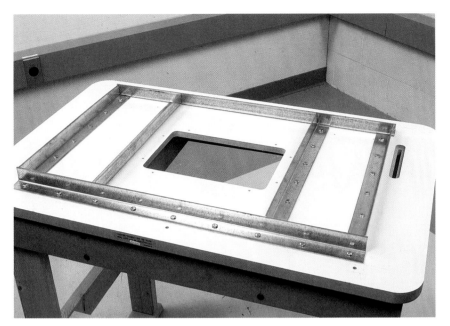

Angle iron reinforces a tabletop without taking up much room beneath the table and usually without interfering with a mounted router.

Making a Metal Table Frame

If you would like to make your own metal table frame, use angle iron (available at most hardware stores) with each leg 1½ in. to 2 in. wide. When buying the angle iron, check it by eye or with a straightedge to make sure it is straight. The metal will cut, drill, and file easily if your tools are sharp, so use a fresh hacksaw blade and drill bit.

For most tables, the layout of the angle iron shown in the photo above will work well. The four pieces making up the perimeter of the frame should be set back no more than 3 in. from the edge of the table. The top, however, won't stay flat without additional reinforcement on each side of the panel cutout. To add as much support as possible where the table is weakened by the miter slot, run the pieces framing the panel opening perpendicular to the miter-gauge groove.

Drill screw holes every 3 in. or so along the length of the angle iron. When laying out the holes, avoid placing a screw directly under the miter groove. Be sure to locate a few screws as close to the miter-gauge slot as possible to reinforce this inherently weak area.

To attach the top to the frame, use #10 or #12 sheet-metal screws that are long enough to penetrate the table by ¾ in. (almost all commercial tops are 1 in. to 1⅛ in. thick). Drill pilot holes the size of the inner diameter of the screw into the underside of the table, making sure you use a stop on the drill to avoid drilling all the way through the tabletop. When tightening down the screws, go easy or they'll strip. The point of using a lot of screws and starting out with a flat table is to avoid placing a lot of stress on any one screw.

Making a Wood Table Frame

To hold the table flat, you must make your wood or MDF frame at least 4 in. deep. Start by laying it out as shown in the photo below, butt-joining the corners of the frame with bolts and barrel nuts to make a simple but very strong structure. Wide-head bolts and barrel nuts (also called cross dowels) are commonly used for knockdown furniture and are available at woodworking stores and through catalogs.

To attach the wood frame to the top, use either flat-headed bolts or knockdown furniture bolts countersunk into the table's surface (see the top right photo on the facing page). Run the bolts into barrel nuts set into the frame members. Because through bolts hold with far greater force than screws, you'll need only a few bolts to attach the top to the frame: one bolt at each corner of the table and one in the middle of the longer frame members.

With the table flattened and reinforced, remount it on its base. Before replacing the drop-in panel, use a fine file to bevel the edge of the laminate around the perimeter of the cutout. The bevel will help to prevent a workpiece from catching on the edge of the opening if the table and drop-in panel aren't quite even with each other (see the top left photo on the facing page).

CHECKING THE DROP-IN PANEL

Next, check the drop-in panel by laying a straightedge across it to see if it is flat (see the bottom left photo on the facing page). If the panel has been

Bolts and barrel nuts designed for knockdown furniture make a reliable joint in MDF, but drilling must be precise so the hardware will line up during assembly.

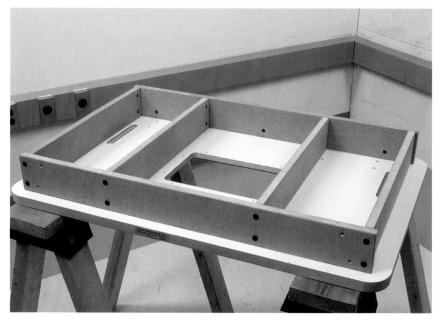

A wood frame needs to be deeper than a metal frame to hold the top flat, but it is easier and less expensive to make.

Beveling the edge of the panel cutout with a file will help stock slide smoothly without catching.

The same type of bolts used to assemble the frame can be countersunk and used to attach the tabletop to wood framework.

pulled down by the weight of the router, you can flatten it as described on pp. 167–168. To prevent the sag from returning, it's a good idea to remove the router and panel from the table between jobs.

The alternative fix for a warped panel is to replace it with one made from a stiffer material. Before making a new panel from scratch, it is worth checking with the table's manufacturer; if there have been a number of similar complaints, a better replacement may already be available.

If you can't buy a precut replacement panel made for your table, several woodworking catalogs sell blank sheets of high-strength plastic for this

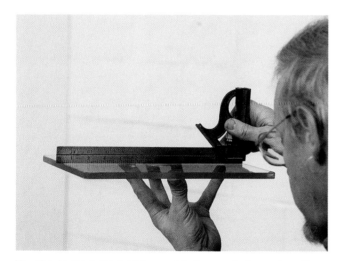

You should check the drop-in panel that supports the router with a straightedge. Panels made from transparent plastic often are bowed or twisted.

A variety of plastics are used for making drop-in panels; the dark-colored phenolics are generally the sturdiest and flattest.

purpose. Plastics are fairly easy to shape using woodworking tools but they dull steel blades quickly, so use carbide-edged tools as much as possible. The easiest way to make a new insert is to attach the slightly oversize blank replacement panel to the original panel with double-sided tape and then trim the new one to size using a bearing-guided straight router bit.

Using a file, slightly bevel the edges of the panel and the router bit opening (see the photo at left). In addition to making it less likely that stock will catch, a bevel also cuts down on burrs forming on the panel's edge that can score the workpiece. If the panel has inserts to give a close fit around the router bit, check that they fit flush with the panel's surface. If the insert is too high, try setting it lower by deepening the rabbet it sits in or by sanding the insert until it is thinner. If the insert is low, use tape shims to raise it. As with the panel and table opening, use a file to bevel the inner and outer edges of the insert.

LEVELING THE PANEL

Most commercially made tables have leveling screws to bring the panel flush with the table surface. If your table doesn't have them, they're easy to add by screwing four wood blocks to the underside of the table at the corners of the cutout (see the photo below). Run fine-thread drywall screws into the blocks from the top; the screw heads will hold the panel at the right height. If the panel sits higher than the table surface, use a router or a chisel to deepen the rabbet.

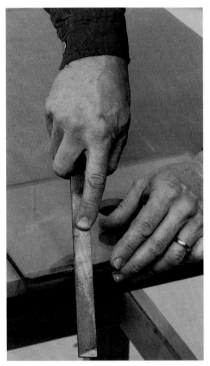

To allow the workpiece to slide smoothly, file a bevel on the edges of the drop-in panel and at the bit opening.

Adding corner blocks with height-adjusting screws on the underside of the table will make it easy to bring the drop-in panel flush with the top.

To check that the panel and table are flush, cut a small hardwood block as shown in the bottom photo. By leaving the short edges on the end of the block sharp, the block will snag on even a slightly raised edge as it slides over the joint where the panel meets the top.

To bring the panel flush with the tabletop, back off all four screws and drop the panel below the table. Next, using the two screws on either cor-

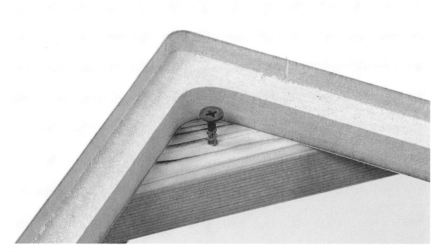

Height-adjustment screws are simply fine-thread drywall screws. Screw heads support the drop-in panel.

A hardwood test block with sharp edges makes it easy to test whether the drop-in panel and table are even with each other.

ner of the infeed side of the panel, raise the panel until the block barely catches on the edge of the panel. With the height of the infeed side roughly set, move to the outfeed edge. The block should catch as you try to slide the block from the panel onto the table. Using the two corner screws on the outfeed side, raise the panel until the block can just slide onto the table without catching.

Now go back to the infeed side of the panel and fine-tune the height. Raise the panel with the infeed side screws in small increments until the block just catches on the panel's edge nearest the screw you're adjusting, then back off the screw slightly so the block slides smoothly from the table onto the panel. Fine-tune both screws on the infeed side of the panel until the block slides smoothly from the table onto the panel all along the infeed edge.

The final step is to adjust the height of the outfeed side of the panel. Working again in small increments, lower the panel height until the block just begins to catch on the table's edge, then raise the panel slightly until the block slides smoothly. Adjust both screws until the block slides without catching.

If the table has additional adjustment screws along the edges of the panel in between the corner screws, adjust these only after you have set the four corner screws to the correct height. Raise the first intermediate screw until the panel starts to rock on that screw, then back it off until the rocking disappears. When the panel no longer rocks, the screw is properly adjusted and you can move on to the next one.

Adjusting the Fence

To work properly, the fence must be straight and square to the table, but unless the tabletop is flat, there will be no reliable reference surface to use in checking it. A warped table also will cause a lightweight fence to twist out of shape each time it is clamped down, making a tune-up pointless. Check and correct any problems with your table, as outlined in the previous section, before checking the table's fence.

Start by clamping the fence over the centerline of the table, with the fence over the bit opening. Using a level as a straightedge, check to see whether the fence is straight along both top and bottom edges (see the photo on the facing page). If the face of the fence is split, look for a step between the two halves where they meet in the middle.

Next, check the fence every 3 in. to 4 in. using a square. It's fairly typical for a fence to be square to the table over half its length and then start to run out over the second half.

The best fix for a fence that is out of line will depend on its facing material. You can run solid wood faces over a jointer to true it. Plastic, plywood, and MDF can't be jointed; you'll have to true them by placing shims between the face and the aluminum angle.

Use a reliable square to check that the fence is square to the table along its entire length.

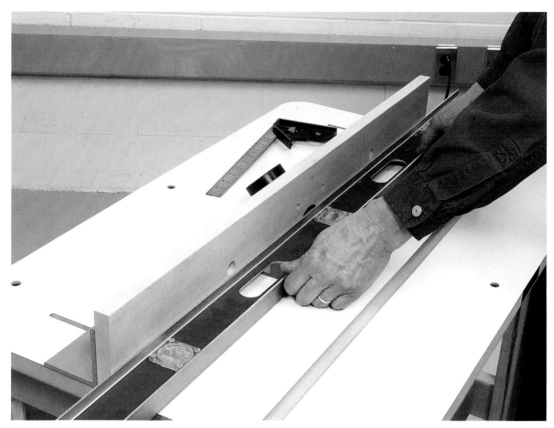

Using a level as a straightedge, check both the top and bottom of the fence for straightness.

TRUING UP A WOOD-FACED FENCE

To true up a fence with a wood face, start by removing the face and checking if it's warped. If the back surface isn't flat, true it up on a jointer. If the face is badly warped, the best approach is to make a new face using a seasoned, stable hardwood, preferably straight grained and quartersawn. If the original fence had a split face, you can replace it with a simpler one-piece face with an opening for the bit.

Try the fit between the trued-up back of the face and the aluminum extrusion while the fence is clamped to the table. If the face doesn't fit well against the frame over its entire length, the frame either has a permanent twist in it or the mounting clamps are bending the frame, possibly because the table isn't flat. If the frame straightens out when the clamps are released, you need to correct a problem with the clamps or the table to eliminate the twist.

If the twist in the aluminum is permanent, either handplane the back of the face to fit, or use shims to achieve a good fit between the face and the frame when they are screwed together. You don't want to twist either the wood or the frame once the fence is assembled.

One advantage of a wood fence is that it can be flattened on a jointer, but make sure the jointer knives will not strike any metal fasteners or hardware.

The final step in truing up a wood-faced fence is to pass the assembled fence over a jointer to flatten and square the front face (see the photo above). Make absolutely sure the hardware used to attach the face is countersunk to prevent damage to the jointer's knives and possible injury to you.

TRUING UP A PLASTIC-, PLYWOOD-, OR MDF-FACED FENCE

Because plastic, plywood, and MDF fences can't be planed flat, you must square them up by using shims. The most convenient material to use to shim the faces is ordinary masking tape or a slightly heavier version, the white artist's mounting tape shown in the photo on the facing page. Duct tape doesn't work very well because it's too soft, creeps under pressure, and is difficult to peel off.

To get the face square to the table, start with the fence clamped down over the middle of the table. If it has a split face design, separate the two halves by an inch or so and tighten up the face attachment bolts. Use a square to determine if the face needs to be shimmed at the top or bottom.

Because plywood, MDF, or plastic faces can't be planed, shim them true with layers of tape applied to the aluminum frame of the fence.

Straightening a Split Fence

The shorter halves of a split fence rarely need to be straightened, but one side may need to be shimmed out to eliminate a step between the two halves.

If the face is out over its full length, remove it and apply a continuous line of tape along the top or bottom edge of the aluminum to correct the problem. Then reattach the face and check it again with a square; a second or third layer of tape may be needed. If the facing is out of square over only part of its length, just apply tape behind the section that is out. On a split fence, adjust the infeed and outfeed sides separately.

Once the face is squared up, use a level to check that it is still straight. If the face needs straightening, add tape. Apply the tape to straighten the face lengthwise on top of any shim layer you added earlier to make the face square to the table.

The Miter Saw

M iter saws were originally designed for carpenters, not cabinet-makers. Early models were not reliably accurate, and they developed a poor reputation among the cabinetmakers who tried to use them for crosscutting. The saws were often relegated to a spot near the lumber rack, where they were used only for cutting stock roughly to length.

The miter saw, however, has since evolved into a tool capable of furniture-grade precision. It is now used for joinery, not just rough work. Its primary advantage is in crosscutting wide or long stock that would be very difficult to handle on a table saw with a miter gauge.

Today's miter saws (above) are precise enough for furniture making. Principal elements of a miter saw (right) include a base with a revolving turntable to control the cutting angle, a fence, and a power head. On compound miter saws, the power head pivots to make bevel cuts.

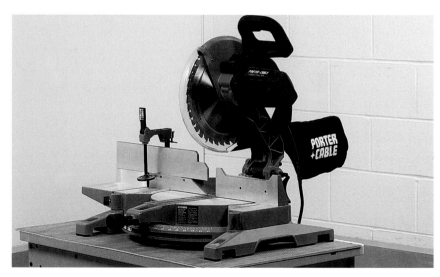

The Miter Saw

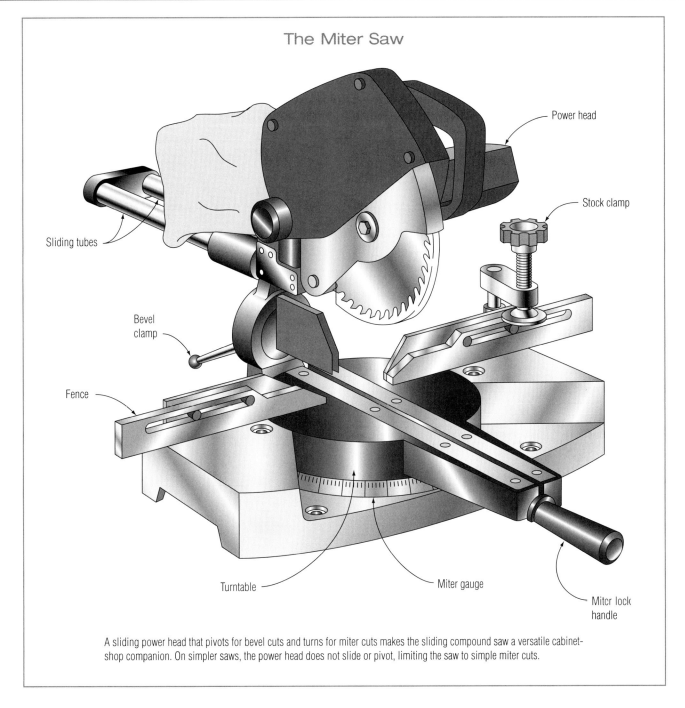

Power head

Stock clamp

Sliding tubes

Bevel clamp

Fence

Turntable

Miter gauge

Miter lock handle

A sliding power head that pivots for bevel cuts and turns for miter cuts makes the sliding compound saw a versatile cabinet-shop companion. On simpler saws, the power head does not slide or pivot, limiting the saw to simple miter cuts.

Anatomy

In its most basic form, the miter saw has a table, a fence, and a power head that pivots downward to make a cut. It is this downward cutting motion that gives the miter saw its common nickname: the chopsaw.

The sliding compound miter saw is a tempting investment. The slide gives extra crosscutting and mitering capacity, but sliding saws are much less accurate and difficult to keep accurate if adjusted. Unless you really need the extra cutting capacity, choose a 10-in. or 12-in. nonsliding saw for cabinet work and invest the money you save in a top-quality blade.

With the power head mounted on tubes, this compound miter saw has a crosscut capacity of 12 in.

The base of the saw is primarily a large turntable flanked by a small fixed table on each side. In the middle of the turntable is a slot that allows the blade to cut all the way through the workpiece without striking any part of the tool. As the turntable is pivoted to adjust the miter angle, the power-head assembly swings with it. The fence is attached to the side tables and stays in a fixed position when the turntable is rotated. On even the simplest saws, the turntable can be swung up to 45 degrees, left or right, to make basic miter cuts. In addition to stops at the 45-degree position, most saws also have stops, or detents, at 22½ degrees and 30 degrees that are used for mitering six- and eight-sided frames.

On some saws, the power head also pivots—up to 45 degrees to the left and occasionally also to the right—to make bevel cuts. Saws with both miter and bevel capacity are called compound miter saws. A saw that can bevel both left and right is called a dual or double compound saw. The ability to cut compound angles is very useful for some jobs—when installing crown molding, for example—but it has limited applications in cabinet work.

The maximum width of stock that can be cut is limited in simple saws by the blade diameter. At 90 degrees, a 10-in. blade can cut stock up to 6 in. wide, whereas a 12-in. saw can handle 8-in.-wide stock. With the turntable rotated to a 45-degree miter, a 10-in. saw is limited to cutting 4-in.-wide stock; a 12-in. saw will typically cut just under 6-in.-wide stock at this angle.

By mounting the power head on a pair of tubes that allow the head to slide forward, the capacity of the saw can be increased, typically to a full 12 in. Almost all sliding saws also have the capability of cutting compound angles. To work smoothly and accurately, the sliding tubes and their bearings must be carefully machined and aligned. Because of their design, sliding saws, no matter how well made, are less accurate and more expensive than a simple miter saw.

Troubleshooting

Problems with miter saws fall into two broad categories: inaccurate cuts that cause out-of-square or gapping joints, and rough surfaces on the cut face of the stock. Inaccurate cuts can usually be traced to an adjustment that's out of whack, but occasionally the problem is caused by worn or loose-fitting parts. To get the saw to cut well, tune it up as outlined in the next section. I've included procedures for checking wear and for making quick adjustments to eliminate excess play in the saw's pivots and slides.

Rough cuts are almost always due to the sawblade. When miter saws come from the factory, they are typically equipped with mediocre-quality 36- or 40-tooth blades meant for trim carpentry. For accurate, glass-smooth cuts, you need to invest in a top-of-the-line 80-tooth blade designed specifically for miter saws.

Setting Up a Miter Saw

To deliver the accuracy required in cabinetmaking, a miter saw should be mounted on a solid stand with stock support tables extending several feet to either side of the blade. Both commercial and shopmade stands come in two basic designs. In one, the side tables come up flush with the table of the miter saw. This type of stand requires leveling screws at the ends of the tables so the surfaces can be aligned.

In the second most common design, the one I prefer, side tables are below the level of the miter saw's turntable. A simple, sturdy version is a one-piece table, long and narrow, with the saw mounted at its midpoint. The table surface should be parallel to the miter saw's table, but it doesn't need to be flush with it. Stock is supported by blocks of wood that make up the height difference between the table and the saw.

The primary advantage of this design is that it allows you to make adjustments easily for bowed stock. For accurate joinery, the workpiece must lie perfectly flat against the miter saw's table and fence. If you are trying to cut a longer piece of stock that has even a slight bow in it, you can use a block to keep the cut end in contact with the table. I keep blocks of different thicknesses on hand for just this situation (see the bottom photo on p. 182).

Use a Miter-Saw Blade

Table-saw blades should never be used on a miter saw. The sharply angled teeth will grab and lift the stock off the miter saw's table and cause uncontrollable kickback.

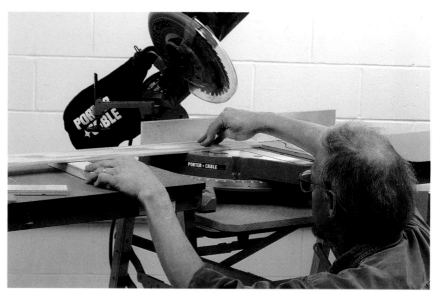

For a precision cut, the stock must sit squarely on the saw's table. Always check that there's a good fit before sawing.

A selection of blocks used to elevate the stock to different heights can be useful. By adjusting the end of a bowed board, the other end can be held flat on the table for the cut.

INSTALLING AN AUXILIARY FENCE

For no obvious reason, the fences on most compound miter saws are very low, sometimes just more than an inch high with a very large gap between the left and right halves. For cabinet work, these fences don't support the stock adequately, since they allow the board to shift or wobble, which can ruin a cut. On most saws it is possible to add a plywood facing to the metal

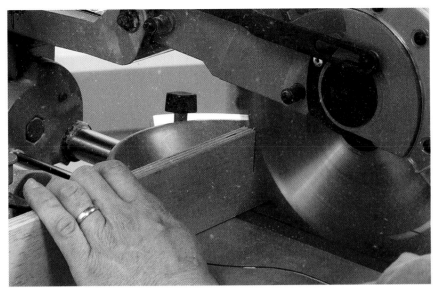

An auxiliary fence is attached as a single piece and then sawn in half. The narrow gap reduces tearout on the back of the workpiece.

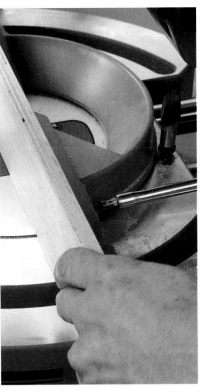

Adding a wider and longer plywood facing to a miter saw's original metal fence will improve the saw's performance.

To provide clearance for the pivoting power head on compound miter saws, the top of the auxiliary fence must be notched.

fence to correct the problem. Many saws have screw holes already drilled in the metal fence to make attaching the auxiliary fence a simple job.

Because the thickness of the facing reduces the cutting capacity of the saw, use ½-in.-thick plywood to minimize the loss. The ideal material is a nine-ply Finnish or Baltic birch plywood that should be free of warp. Make the add-on fence from a single piece of plywood, 3 in. to 4 in. tall and a couple of inches longer than the miter saw's fence. Attach the facing to the metal fence with sheet-metal screws from behind or countersunk flat-head machine screws and nuts that go through the plywood.

Once the facing is attached, cut through it with the saw set for a square cut and then left and right 45-degree miters. If you don't use a compound saw's bevel function, there is no need to cut away the facing with the power head tipped over. By not cutting away a large wedge of the facing for bevel cuts, the facing will better support the stock for regular mitering. If you do use the saw for bevel cutting, it will be necessary to notch the facing's top edge to clear the saw's frame when it is tipped over.

Tuning Up a Miter Saw

To tune up your saw, start by removing dust and chips from the base. You can safely use compressed air to blow out the crevices under the saw's table, but don't use high-pressure air to clean off the tubes of sliding saws because the blast of air can force dust past seals that protect the bearings. Remove the blade and clean out the inside of the blade housing and the guard because pitch often builds up in these areas.

Dust, especially if it contains pitch, also may accumulate on the sliding tubes and interfere with the smooth motion of the saw. To remove it,

For smooth operation, the tubes on a sliding saw must be kept clean. Penetrating oil is an effective solvent for gum and pitch.

You should occasionally use light machine oil to lubricate the moving parts of the saw.

The pivot bolt for the saw's turntable must be tightened carefully to eliminate any free play while still allowing the turntable to rotate smoothly.

If the saw has an adjustable power head tilt bolt, you should check and tighten it to eliminate any excess motion.

spray some penetrating oil, which is an efficient pitch solvent, on a clean towel and use it to wipe down the tubes (see the photo at left on the facing page). Don't leave oil on the tubes; wipe them down with a dry towel once they are clean.

Using a light machine oil, lubricate the plunge pivot bolt that the head rotates on when the saw is pulled downward. On many saws, most of the remaining pivots and lock components cannot be reached for lubrication, but it is worth lubricating those that are accessible. Components that are difficult to move or grate when they're turned should be disassembled, if needed, for lubrication.

CHECKING FOR EXCESSIVE PLAY

If there is play in the saw, you should track it down and eliminate it before adjusting the fence and stops for squareness. Start by making sure the pivot bolts for the turntable and head tilt are as snug as possible while still allowing the parts to rotate smoothly. Beginning with the turntable, gradually tighten the pivot bolt on the underside of the table until the table starts to drag when you turn it, then back off the bolt until the drag just disappears (see the photo at left above). Repeat the procedure with the head tilt bolt on the back of the saw.

Loose bearings on a saw's sliding tubes can cause inaccurate cuts. You should check and tighten bearings as part of a tune-up.

Once you've tightened the saw's bearings and pivot bolts, there should be no play when side pressure is applied to the power head.

On sliding saws, there may be one or more bolts that can be adjusted to take up play in the bushings of the sliding tubes. Adjusting these screws properly can be tricky, so it's a good idea to follow the instructions in the saw's manual. Lacking specific instructions, gradually and evenly tighten all the screws for one bar until they all drag slightly. Back the screws off to eliminate the drag, then lock the screws and repeat the procedure for the second tube. On some saws, the plunge pivot bolt can be adjusted. If so, tighten the bolt until you can feel resistance when you pivot the saw downward and then back off and lock the bolt.

With all the slack taken out of the pivot bolts and the sliding tubes, it is time to check the saw for any other sources of play. Clamp or bolt the miter saw onto a sturdy benchtop if it is not already attached to a stand. For the first check, lock the saw for a square cut and grasp the power head from the side with the saw in the raised position. On sliding saws, position the head at either the beginning or the end of the stroke, choosing the position that minimizes the projection of the tubes. Try to flex the saw by pushing and pulling against the head (see the photo at left). Any play here is probably due to a worn plunge pivot bolt or its bushings. Worn bushings on the sliding tubes also are a possible cause.

Next, check the detent mechanism on the turntable by rotating the turntable away from and then back to 90 degrees. The detent should engage the notch crisply at the 90-degree position. Using the handle on the turntable, try to shift it left and right. The detent should hold the table solidly without having to engage the backup lock. If the turntable doesn't hold solidly, check the detent mechanism because sawdust or wear

The depth of cut should be adjusted so the blade enters the slot to the bottom edge of the gullets between teeth. After making this adjustment, check that the sawblade won't strike the bottom or ends of the slot.

may prevent the locating pin from seating fully. A worn or loose center pivot bolt on the turntable also can allow play. Repeat this test at the left and right 45-degree miter positions.

With the turntable lock engaged, pull the saw down to the cutting position and try to push it to the left and right with moderate force. Except for some flexing on sliding saws, the power head should not shift. Play here is most likely due to a worn plunge pivot bolt or worn bushings on the sliding tubes, but a worn or loose center bolt on the turntable is another possible source of trouble.

ADJUSTING THE BLADE DEPTH STOP

Press down on the power head until it rests against the depth stop. Holding the saw down, look at the side of the blade to check the depth setting. On most saws, the blade should enter the table slot to the bottom of the gullets on the blade's teeth. If the depth needs adjusting, you'll find the stop bolt on the saw frame close to the plunge pivot bolt (see the photo above). If you increase the depth of cut, make sure the blade can't come into contact with the turntable or the pivot assembly.

CHECKING AND ADJUSTING THE BLADE GUARD

You should check the linkage of the blade guard frequently to make sure it operates smoothly. The guard should fully enclose the blade when the power head is raised. Some saws come with poorly designed mechanisms hidden inside the blade housing that gum up and are hard to service. You should clean and lubricate the guard's linkage and check that it does not stick as it pivots (see the top photo on p. 188). Bending the linkage's main arm slightly may help to eliminate jamming or sticking. On most saws, the pin that engages the linkage can be adjusted to make the guard pivot back just as it reaches the stock (see the bottom photo).

A sticking guard that doesn't close fully is very dangerous. It should be free to move and fully enclose the blade when the power head is raised after a cut.

The position of the pin that retracts the blade guard should be adjusted so it lifts the guard just as it reaches the stock.

CHECKING THE FENCE

The fence on a miter saw is typically an aluminum casting with a large, semicircular segment connecting the left and right halves. The looped part of the casting is a weak spot that can get bent, throwing the two faces of the fence out of line. A straightedge placed across the fence should line up perfectly with each side. If not, you have two options: Try to straighten the casting, or add facings to the fence and shim them until they are aligned.

To straighten the casting, remove the fence from the saw and set it up across two blocks on the benchtop. Press down on the high side of the fence with moderate pressure for just a moment, then recheck it for straightness. Don't use a lot of pressure, at least not for the first try. Most fences are not very stiff and you can easily overdo it.

The basic procedure for adding facings is explained on pp. 182–184. You can add shims after separating the two halves of the fence with a saw cut.

SQUARING UP THE VERTICAL ALIGNMENT

The most accurate way to adjust the various settings on a miter saw is to make test cuts and then check the end of the test board with an accurate square. An alternate method is to use a square to check the alignment of the sawblade to the table or fence. This technique, frequently recommended in manuals, is invariably less accurate. The critical point is to get a square end on the board, so you might as well measure that directly.

On a compound-angle saw, start by checking the saw for square vertically. You will need two test boards ¾ in. thick, 2 in. to 3 in. wide, and 2 ft. long. One board is required for the vertical test and the second one will be needed later for setting the 45-degree miter stop. Using a jointer and a

Shimming Facing Material

Masking tape makes a convenient shim material. It is thin enough for making small adjustments, and it will stay in place as you screw the facing onto the original fence.

Make a test cut to check the vertical alignment of the blade. The mark on the board's face designates the good edge of the board, which should be placed against the table.

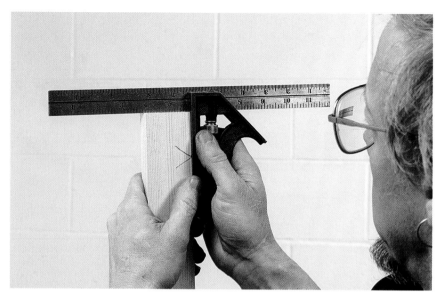

To check the accuracy of the cut, place the head of the square against the designated good edge of the test piece.

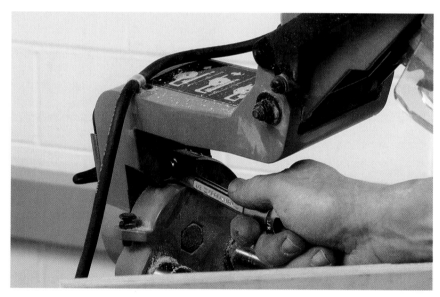

You can bring an out-of-square saw into alignment by adjusting a stop bolt on the back of the tool.

planer, flatten and square the edges of the boards. For accuracy, it is especially critical that one edge of each board be dead straight. This edge, which goes against the miter saw's table or fence when making a test cut, should be designated with a V mark placed on an adjoining wide face of the board.

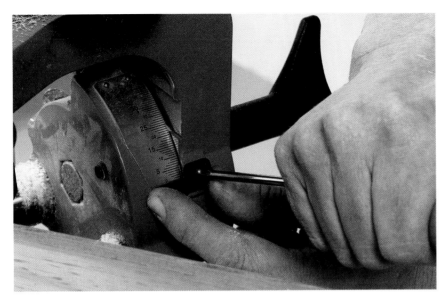

Once the saw is squared up vertically, set the pointer on the tilt scale to 0 degrees. You can bend the pointer to bring it closer to the scale.

To test the saw's vertical adjustment, set and lock the saw at both the vertical and horizontal 90-degree stops. Stand one of the test boards on edge against the fence, with the designated good edge down, and make a test cut (see the photo on p. 189). Next, using a reliable square and holding the board upright in front of a bright background, check for an exact 90-degree cut. If the saw is set properly, the blade of the square will fit tightly against the board's end (see the top photo on the facing page). For a true reading when using the square, be sure to place the tool's stock against the designated edge, the edge that was against the saw's table when you made the test cut.

If the cut end isn't square, adjusting the stop is usually simple. Ordinarily, the stop is an easily reached bolt and locknut in the tilt assembly on the back of the saw. Some saws come with a special wrench to loosen the locknut because clearances can be tight. To make the adjustment, tilt the power head to take pressure off the stop bolt and then loosen its locknut (see the bottom photo on the facing page). Turn the bolt to correct the misalignment, lock it down again, and take another test cut. It may take a couple of tries to get a perfectly square cut on the test piece.

Once you have the saw cutting a square corner, set the pointer on the bevel angle scale to line up precisely with the scale's 0-degree mark (see the photo above). On many saws, the angle will be easier to read and the positioning more precise if the pointer is bent to sit as close as possible to the scale without actually touching it.

Make a test cut to check the saw's miter accuracy, placing the designated good edge of the board against the fence.

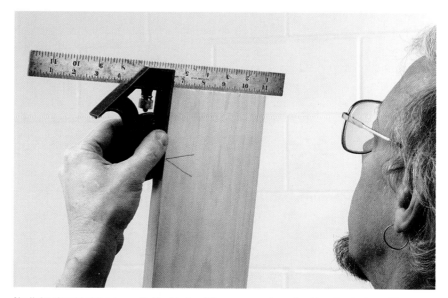

No light should shine beneath the blade of the square when checking the test cut. The 90-degree crosscut is the single most critical adjustment on the saw.

SQUARING UP THE HORIZONTAL ALIGNMENT

To check a saw's horizontal, or miter, alignment, make a test board ¾ in. thick, 3 in. to 4 in. wide, and a couple of feet long. If you also plan to check the saw's 45-degree miter detents, make up two boards of this size. The boards should be jointed and planed square and have a good edge marked.

Before shifting the fence to square up the saw, mark its starting position on the table with a scribe or pencil line.

Square up the fence horizontally by loosening the mounting bolts and shifting it slightly. Scribe marks on the table make it easy to judge how far you have shifted the fence.

Begin by locking the saw against both the vertical and horizontal 90-degree stops. Place the board on the table with its designated edge against the fence and take a test cut, then hold the board in front of a bright background and check the cut end with a good square (see the bottom photo on the facing page). Remember to place the head of the square against the designated edge of the board. If the horizontal alignment is off, it is corrected by shifting the fence. Before loosening the bolts that clamp the fence to the table, make a short scribe or pencil line at each end of the fence to mark its starting position as shown in the photo above. The marks will make it easier to see how much you have shifted the fence as you square it up.

Loosen all the bolts holding the fence and check that no sawdust is trapped between the bottom flange of the fence and the saw's table. Shift the fence to correct the alignment, using the scribe marks to guide you, and retighten the bolts (see the top photo at right). Continue to make test cuts and adjustments until the setting is correct. If the saw is properly aligned, it should give a square cut with the stock against either the left- or right-hand fences. If this is not the case, the fence must be bent out of alignment and should be rechecked.

Once the saw is cutting a square corner, set the pointer on the miter-angle scale to line up precisely with the scale's 0-degree mark. The pointer should be bent so it just clears the scale.

Once the saw is adjusted, make sure the pointer is aligned with the 0-degree setting on the turntable.

Cutting the ends of two carefully prepared test boards will help determine whether the 45-degree bevel angle setting is correct.

The most accurate test of the saw's bevel adjustment is to assemble a joint made from two test boards and check whether the corner is square.

ADJUSTING 45-DEGREE VERTICAL ALIGNMENT

With the vertical and horizontal square stops set, you can now set the stop for tilting the head 45 degrees to the left. For this test, you will need two boards, both 2 in. to 3 in. wide and a couple of feet long. Square up the boards and mark the best edge on each board with a V on the adjacent face.

Rotate the saw against the 45-degree vertical stop, lock it there, then take a miter cut off the ends of both boards, keeping the designated edges against the table. Assemble the joint on a flat benchtop and check that the corner is square (see the bottom photo on the facing page). If the joint isn't square, adjust the stop bolt and make another test cut. It probably will take several attempts to get a perfectly square corner.

CHECKING 45-DEGREE HORIZONTAL ALIGNMENT

To run this check, you will again need two boards that are ¾ in. thick, 3 in. to 4 in. wide, and a couple of feet long prepared in the same way as the other test pieces. This is only a check of the saw's 45-degree left and right stops; the stops can't be adjusted. The saw's turntable is positioned by a pin that engages a series of notches machined into a large casting under the table. The notches can't be repositioned. If they weren't manufactured correctly, you are out of luck.

The overall alignment of the turntable was set when you adjusted the position of the fence for cutting a square end on a test board. For this reason, the straightness of the fence and the precision of a square cut should always be checked and adjusted before checking the 45-degree stops.

There are three steps involved in this test: a cut of both boards with the saw swung to the left, a cut of both with the saw to the right, and finally one cut on the left and one cut on the right to make a square corner. This last test is the most critical because that's the way most miter joints are made in the shop.

Adjust the stop bolt for the vertical 45-degree setting. Because of tight clearances around the bolt, use the wrench supplied with the saw to loosen the locknut.

Make a test cut to check the turntable's 45-degree setting with the designated edge of the test piece against the saw's fence.

Realigning the Fence

If a 45-degree stop is off, you could realign the fence to correct the problem, but the square cut would then be thrown off. If your primary use of the saw is for cutting 45-degree miters, realigning the fence may be a sensible option, but there is still no guarantee that both the left and right 45-degree positions will be correct at the same time.

After sawing both boards, assemble the test joint and check it for square.

If a saw can't be adjusted to cut correctly in both the straight and 45-degree positions, use a shim to compensate when making the miter cuts.

Other than the position of the turntable, all three tests are run the same way. Cut the boards after placing them flat on the table, with the best edge against the fence as shown in the bottom photo on p. 195. After sawing, butt the cut ends together on a flat surface and check the resulting corner with a good square (see the top photo on the facing page). Ideally, you'll find that all three combinations work, each producing 90-degree corners. If the joint made with one board cut on each side is good but the joints made with both boards cut on either the left or the right are off, chances are good that the fence is out of alignment. You should go back and recheck the setting for the horizontal square cut.

If the saw can't produce good miter joints in all combinations of left and right cuts, you should remove the saw's turntable and examine the alignment notches. With luck, you will find sawdust packed in one or more of the notches and the problem will be solved with a simple cleaning. If that doesn't solve the problem and the machine is still under warranty, contact the manufacturer. Another option, as I explained earlier, is to align the fence to create a good joint on one side of the blade at the loss of accuracy in other turntable positions.

A final fix, and one I have used often when all else has failed, is to find by trial and error a shim size that when slipped between the fence and the stock corrects the problem and creates a square corner (see the bottom photo on the facing page). You would use the shim when making miter cuts but remove it when making square cuts.

Sources

Bridge City Tool Works, Inc.
5820 N.E. Hassalo
Portland, OR 97213
(800) 253-3332
Measuring tools

**Delta International Machinery
Corp.** (see Porter Cable/Delta)

DeWALT® Power Tools
701 E. Joppa Rd.
Towson, MD 21286
(800) 433-9258

Dremel®
4915 21st St.
Racine, WI 53406
(800) 437-3635
Rotary and grinding tools

Emerson Tool Company™
8100 W. Florissant
St. Louis, MO 63136
(800) 474-3443
Ridgid power tools

Epic Machinery Group
5128 Westinghouse Blvd.
Charlotte, NC 28273
(704) 588-6627

Fein Tools
1030 Alcon St.
Pittsburgh, PA 15220
(412) 922-8886

FELDER USA
1851 Enterprise Blvd.
West Sacremento, CA 95691
(919) 375-3190

Fenner Drives
311 W. Stiegel St.
Manheim, PA 17545
(800) 243-3374
PowerTwist Plus® V-Belts

Festool (formerly Festo)
Tooltechnic Systems LLC
1187 Coast Village Rd., Ste. 1215
Santa Barbara, CA 93108
(888) 463-3786

Freud® USA
P.O. Box 7187
High Point, NC 27264
(800) 334-4107

Garrett Wade
161 Avenue of the Americas
New York, NY 10013
(800) 221-2942
*Inca machinery; measuring and
hand tools*

General® Mfg Co. Ltd.
835 Cherrier St.
Drummondville, QB J2B 5A8
Canada
(819) 472-1161

Hartville Tool
13163 Market Ave. N.
Hartville, OH 44632
(800) 345-2396
Measuring and setup tools

Highland Hardware
1045 N. Highland Ave. N.E.
Atlanta, GA 30306-3592
(800) 241-6784
Measuring and hand tools

Hitachi Power Tools
3950 Steve Reynolds Blvd.
Norcross, GA 30093
(800) 829-4752

Injecta Machinery/Eagle Tools
2217 El Sol Ave.
Altadena, CA 91001
(800) 203-0023
Inca machinery

Iturra Design
4636 Fulton Rd.
Jacksonville, FL 32225
(888) 722-7078
Repair and upgrade parts for bandsaws

Jet Equipment & Tools
P.O. Box 1937
Auburn, WA 98071-1937
(800) 274-6848

Lee Valley Tools, Ltd.
1090 Morrison Dr.
Ottawa, ON K2H 1C2
Canada
(613) 596-0350
Measuring and hand tools

Makita® USA, Inc.
14930 Northam St.
La Mirada, CA 90638
(714) 522-8088

Milwaukee® Electric Tool Co.
13135 W. Lisbon Rd.
Brookfield, WI 53005
(262) 781-3600

MSC Industrial Supply
75 Maxess Rd.
Melville, NY 11747
(800) 645-7270
Measuring tools and machine parts

Panasonic® Personal and Professional Products
One Panasonic Way 4A-3
Secaucus, NJ 07094
(201) 271-3476

Porter Cable/Delta
4825 Hwy. 45 N.
Jackson, TN 38302
(888) 848-5175

Rockler™ Woodworking and Hardware
4365 Willow Dr.
Medina, MN 55340
(800) 279-4441
Measuring and hand tools

S-B Power Tool Company
4300 W. Peterson Ave.
Chicago, IL 60646
(877) 754-5990 (Skil)
(877) 267-2499 (Bosch)
Skil power tools; Bosch power tools

Sears™ Power and Hand Tools
P.O. Box 14588
Des Moines, IA 50306-3588
(800) 290-1245
Craftsman power tools

Sunhill Machinery™
500 Andover Park E.
Seattle, WA 98188
(800) 929-4321
Co-matic power feeders; Sheng Shing wide-belt sanders

Woodcraft®
560 Airport Industrial Park
P.O. Box 1686
Parkersburg, WV 26102
(800) 225-1153
Measuring and hand tools

Woodworker's Supply, Inc.®
1108 N. Glenn Rd.
Casper, WY 82601
(800) 645-9292

Index

NOTE: page references in *italics* indicate a photograph; references in **bold** indicate an illustration.